Studia Fennica
Historica 4

The Finnish Literature Society was founded in 1831 and has from the very beginning engaged in publishing. It nowadays publishes literature in the fields of ethnology and folkloristics, linguistics, literary research and cultural history.
The first volume of Studia Fennica series appeared in 1933.
Since 1992 the series has been divided into three thematic subseries: Ethnologica, Folkloristica and Linguistica. Two additional subseries were formed in 2002, Historica and Litteraria.
In addition to its publishing activities the Finnish Literature Society maintains a folklore archive, a literature archive and a library.

Editorial office
Hallituskatu 1
FIN-00170 Helsinki

The Landscape of Food

The Food Relationship of Town and Country in Modern Times

Edited by Marjatta Hietala – Tanja Vahtikari

Finnish Literature Society • Helsinki

ISBN 951-746-478-9
ISSN 1458-526x

Tammer-Paino Oy,
Tampere 2003

Contents

Contributors

Virginie Amilien is a Senior Researcher at the National Institute for Consumer Research (SIFO), Lysaker, Norway.

P. J. Atkins is Head of Department in the Department of Geography at the University of Durham, United Kingdom.

John Burnett is Emeritus Professor of Social History at Brunel University, United Kingdom.

Maria Concetta Dentoni is a Lecturer in the Department of History at the University of Cagliari, Italy.

Maja Godina Golija is a researcher at the Institute of Slovene Ethnology, Ljubljana, Slovenia.

Adel den Hartog was formerly Senior Lecturer in food habits in the Department of Human Nutrition at Wageningen University and Research Centre WUR, the Netherlands.

Marjatta Hietala is Academy Professor, Professor of General History at the University of Tampere, Finland.

Heather Holmes is a policy analyst in Animal Health and Welfare, Scottish Executive Environment and Rural Affairs Department, Edinburgh, Scotland.

Annemarie de Knecht-van Eekelen is a Head of International Consultancy & Training CITO group, Arnhem, the Netherlands.

Peter A. Koolmees is a Lecturer in the Department of the Science of Food of Animal Origin at Utrecht University, the Netherlands.

Aura Korppi-Tommola is Executive Director of the Federation of Finnish Learned Societies.

Beat Kümin is a Lecturer in early modern European history at Warwick University, United Kingdom.

Derek Oddy is Emeritus Professor of Economic and Social History at the University of Westminster, United Kingdom.

Lydia Sapounaki-Dracaki is Associate Professor at Panteion University, Athens, Greece.

Hans-Jürgen Teuteberg is Emeritus Professor of History at the University of Münster, Germany.

Tanja Vahtikari was a member of the organizing committee of the Sixth Sympsium of ICREFH held in Tampere in 1999. At present she is working on her dissertation on urban history in the Department of History at the University of Tampere, Finland.

Acknowledgements

The editors wish to express their deepest gratitude to Professor Derek Oddy for everything; his contribution has been invaluable in making this volume. We also received much-appreciated assistance from Professor John Burnett and the Members of the Scientific Committee of ICREFH. Moreover, special thanks are due to two anonymous referees for their insightful comments. Finally we wish to thank the Finnish Literature Society for including this volume in its publication series and in particular Rauno Endén, Managing Editor, and Johanna Ilmakunnas and Pauliina Rihto, Editors, for their co-operation.

MARJATTA HIETALA AND TANJA VAHTIKARI

1 Introduction

Interest in food-related issues has been increasing now that urban consumers in developed economies are able to pick and choose what they wish to eat more than ever before. At the same time, though, ordinary townspeople have become alienated from the actual 'sources' of food provisioning; very little of it is visible in western cities. This provides a striking difference from the situation in earlier times when, for example, towns accommodated many animals, especially horses and cows, and slaughterhouses were part of the urban environment. Within the last quarter of the twentieth century, increasing wealth and improved food technology have led to food being supplied from all over the world and increasingly being consumed out-of-season. However, a certain fear of food production systems has developed, particularly to the genetic modification of food crops and the possible effects of animal disease upon man; e.g. Salmonella, foot and mouth disease, Bovine Spongiform Encephalopathy (BSE) and the possibility that this is related to the human Creutzfeldt-Jakob Disease (CJD). Today, urban and rural lifestyles seem to be diverging from each other.

The Sixth Symposium of the International Commission for Research into European Food History (ICREFH)[1] which was held at the University of Tampere, Finland, in 1999, focused its discussions on the relationship between town and country with regard to food provisioning and food consumption. The theme of the symposium, *The landscape of food: town, countryside and the food relationships in modern times,* arose from the scientific debates of previous conferences organized by the Commission dealing with the social history of food and nutrition in Europe. Since its inaugural meeting held in Münster, in 1989, the Commission has concerned itself with a variety of issues in connexion with the food history of Europe in the nineteenth and twentieth centuries, such as the development of European food policies, the relationship of technology and science to food, food and material culture and the health implications of eating and drinking. The Commission explores the issues in food history from an interdisciplinary perspective that includes among others history, ethnology, sociology, geography and economics. The different points of departure of the participants create a diversified approach to the history of food. In the Symposium at the University of Tampere, these converged first through the development of food provisioning (Feeding the towns), then through the reciprocating relationship between the town and rural surroundings (The urban-rural interface) and finally through an examination of the forces of change (Tradition, crisis and innovation). These form the sections of this book.

Although the early towns had fields and gardens for food production, it was evident from the outset that such densely populated settlements were unable to feed themselves solely from the land within their boundaries. Other forms of food provisioning had to be developed. During the Middle Ages and early modern period, in addition to the small gardens of their own, townspeople bought food supplies from farmers on daily markets. Sometimes a city dweller established a special relationship with one or two farmers in the neighbouring countryside who provided him with milk, meat or vegetables. As Europe experienced the urban growth from the nineteenth century onwards, the problem of feeding townspeople was intensified. In the densely structured urban areas there was much less room for gardens. The garden plots that had emerged on the outskirts of towns were ravaged when the town walls were pulled down in many European cities. Also the delivery of foodstuffs from the surrounding countryside to the growing town population was insufficient: short-distance traffic between countryside and town that relied on horses and carts served a limited area. The spread of railways throughout Europe from the 1830s onwards stimulated the movement of people, goods and raw materials and facilitated the maintenance of food supplies to the cities. Railways were important in mobilizing industry, extending urban growth and dispersing urban ways of life into the countryside. These improvements had an impact particularly on the availability of fresh vegetables.

The essays by Hans Teuteberg, Adel den Hartog and Maja Godina Golija in this volume explore the provisioning and consumption of vegetables in European towns in the nineteenth and twentieth centuries. Hans Teuteberg sheds light on the developments in Germany – he concludes that increased consumption of greens in the second half of the nineteenth century resulted not only from urbanization but also from the liberalization of markets and the rationalization of traditional gardening together with improved means of transport. The central tenet in the essay by Maja Godina Golija is that between 1850 and 1950 the Slovenian diet remained remarkably similar in both urban and rural areas, i.e. based on vegetables and cereal dishes. Social distinctions have been the decisive factor influencing vegetable consumption in European cities; in Slovenia, for example, the diffusion of the rarer and more expensive vegetables, such as artichoke, eggplant or asparagus from more affluent households to poorer ones took place only in the 1970s and 1980s. It is interesting, therefore, to cogitate on how different species have spread and which other species they have displaced. Adel den Hartog's essay focuses on Dutch vegetable consumption but also makes a valuable comparison between the present consumption patterns in several European countries, which shows that in southern Europe today, vegetable consumption is considerably higher than in the North.

In Europe, active public promotion of nutrition education, such as the introduction of the so-called food circle – comprising five, six or seven segments depending on the emphasis at a national level – have had a major effect on the increase in the consumption of vegetables, as well as other foods. In this volume these campaigns are discussed through the examples of Finland and the Netherlands by Marjatta Hietala and Adel den Hartog respectively. After the discovery of vitamins in the carly twentieth century and the realization of the fact that vegetables and fruits were good sources of these healthy qualities, it was evident that the diet should always include vegetables. In the Netherlands, as a result of active public promotion and education campaigns through meetings, exhibitions, films, television programmes, newspaper articles, posters and free recipe booklets, urban consumers modified their diet to include a broader variety of vegetables. How successful these campaigns really were and still are remains a fruitful research question. In addition to especially designed campaigns, cookery books, periodicals and magazines have had a role as intermediaries between the newest trends related to food and general usage. Lydia Sapounaki-Dracaki suggests that at the beginning of the twentieth century the 'second generation' of cookery books in Greece adapted recipes of foreign origin, introduced in the 'first generation' cookery books, to Greek taste.

As towns grew larger in the nineteenth and early twentieth centuries, the area required to supply food for them widened which, in part, contributed to the deterioration of food quality. By the 1880s there was European-wide concern over food standards. This challenge was met with the help of new theories on bacteriology and hygienic awareness that gave a basis for the regulation of food quality. This could be seen above all in the founding of market-halls which kept cold, heat, rain and dust away, in regulations for selling milk and meat, and also in founding of food inspectorates. The essay by Peter Koolmees explores the role of veterinarians in safeguarding food quality in Dutch towns. Koolmees suggests that a group of veterinarians contributed significantly to the scientific development of meat inspection. Public abattoirs, in which regular meat inspection took place, were an important part of the public health infrastructure which was built at the end of the nineteenth and early twentieth century in most European countries – usually developing first in the biggest cities.

As the percentage of people in Europe engaged in industrial employment rose during the nineteenth century, there were always fears of food shortages, especially for those in towns. These could occur from a variety of causes. The American Civil War caused distress in 1862–1863 in the cotton towns in Britain; Paris also experienced severe food shortages due to war in 1870–1871. Such fears brought political pressure to bear on countryside interests. This culminated in the free-trade policy adopted across Europe after 1860, which encouraged imports of foodstuffs and raw materials and the removal of trade barriers but, after 1890, tariffs began to be re-imposed. In some states direct pressure was put on the countryside to supply food to the towns: in Russia in the late nineteenth century, food shortages were apparent in rural areas, as the state forced grain to be sent to towns.

In addition to the liberalization of trade, the development of transport had a significant role in the emergence of the global foodstuffs market. As sailing ships were supplemented and ultimately replaced by mechanically-driven vessels, food from North America and countries of the southern hemisphere began to enter Europe and compete with home-grown produce. Although durable foods – bread-grains, rice, tea and coffee – benefited first from long-distance trade, by the last quarter of the nineteenth century improved preservation techniques were important. Frozen beef from the United States arrived in London in 1875 and the first frozen meat from South America reached France in 1878. Tropical fruit could ripen on the voyage but by 1900 it was being stored in low-temperature conditions on board ship. By the end of the nineteenth century many western European ports were equipped with refrigerated storage capacity for imported foodstuffs. Not all regions in Europe benefited equally from these improvements in transport; some areas might still experience local famines. For instance, in 1866–1868, Finland suffered a severe famine that was experienced throughout the country. In some countries, for example in Italy, as the essay by Maria Dentoni demonstrates, the First World War provided a stimulus for the more extensive import of frozen foodstuffs.

The essay by Derek Oddy illustrates how the urban diet in Britain in the later nineteenth century came to depend on processing on an industrial scale in the towns themselves in addition to an increase in foreign imports. In the process, the importance of the countryside declined as a source of food. However, the role of urban food production should not be underestimated, even in the most urbanized societies in Europe. Peter Atkins argues that in British cities of the later nineteenth century, horticulture and milk production were still evident but that this source of food declined steadily in the twentieth century because of competition from rural and foreign producers and the regulatory framework set up by the urban authorities.

The two World Wars of the twentieth century re-emphasized the importance of the countryside as a source of foodstuffs as blockades and embargoes restricted international food trade. The effects of war on the food supply were especially pronounced in towns. As a result there were calls for increased productivity in food production, self-sufficiency, and substitute foodstuffs, such as wild fruits and plants, berries and mushrooms. In her essay, Heather Holmes discusses how the quest for self-sufficiency made government departments in Scotland create special schemes for collecting wild fruits and plants during the Second World War. In addition to providing food, these schemes created a new form of co-operation between urban and rural areas. Often the wartime co-operation in terms of food provisioning developed even more direct forms – townspeople made trips to the surrounding countryside to buy food directly from farmers, as the essay by Aura Korppi-Tommola, which explores the Finnish wartime diet, shows.

There were divergences between regions in individual countries and between the European countries regarding which kinds of 'natural' foodstuffs were mostly utilized. The Germans, for example, rarely used mushrooms to supplement their diet until recently, while for Finns mushrooms have traditionally been part of their diet during both wartime and peace. In eastern Finland, however, these types of foods were much more widely consumed than in the western parts. In her essay, Marjatta Hietala discusses how picking berries and mushrooms is still an important source of income for Finns, especially for those living in the eastern or northern parts of the country.

Urban food provisioning could be manifest in several different ways ranging from travelling tradespeople and open-air markets to market-halls and corner shops. The essay by John Burnett demonstrates that in England this commercialization of people's diet was not limited to urban areas but took place in the rural villages as well. Starting in the seventeenth and eighteenth centuries and accelerating during the nineteenth century, large farms were formed whose ownership was increasingly concentrated into fewer hands. The age of self-sufficiency, therefore, existed mostly only in the nostalgic reminiscences of people living in villages; for example already in 1783 there was at least one licensed tea dealer to every 180 people in England and Scotland.

Inns, taverns and alehouses, commonly thought to be places of sociability and popular culture, were also additionally important centres for the production and consumption of food. The essay by Beat Kümin illustrates that during the early modern period public houses provided people – visitors as well as locals – with ready-made meals and uncooked foods to take away. For the purposes of providing victuals, the keepers of public houses had pieces of land, meadows or gardens. Present-day restaurants wishing to be trendy may well aspire to create tradition; for example, in Norway many restaurants rely on the traditional Norwegian food in their marketing. As the essay by Virginie Amilien suggests, the reality, however, can be different; while words nature and tradition in the restaurants' information leaflets refer to survival and fresh food, menus and cookery are actually far from these folkloric notions.

Perhaps the most wide-ranging change in food production in the twentieth century has been the mechanization of farming. Through rationalization, farming has been turned into an industry. As a counter-reaction vegetarianism and organic farming have gained ground and some urban consumers have again begun to buy their food directly from producers, or to grow some of it themselves, on allotments. In her study of the Netherlands, Annemarie de Knecht-van Eekelen demonstrates the changing functions of urban allotments. Originally, at the end of the nineteenth century, allotments provided food for poor labourers but their use has since developed as a leisure activity. In modern society the use of allotments has become a multi-functional pursuit depending on who is gardening: for some, the *jardin ouvriers* are still utility gardens; for others, they may be recreation areas; and for many a combination of these two functions.

People who left the rural areas and moved into the extensively growing cities in the nineteenth century had to face a drastically different environment that demanded from them many practical and psychological adjustments. Thus, it is no wonder then that in cities "a yearning for country life" developed. Living in the country was considered 'good' and 'healthy' as opposed to 'bad' and 'unhealthy' urban life, regardless of severe agricultural depressions or the poor working conditions of rural labourers. Partially, at least, the myth of a 'better rural life' still exists.[2] However, during the entire period of urbanization there have been attempts not just to return to the countryside but also to improve the quality of town life by bringing nature back to the city. From the nineteenth century onwards, the concept of *rus in urbe* inspired planners to create a more pleasant urban environment also through the creation of urban parks and gardens. From this sprang also the suburban development of the twentieth century, with its allotments. The extensive urbanization of Europe, however, has led the inhabitants of the twenty-first century city to view the countryside as a 'green' recreation area rather than the source of food.

NOTES

1. The International Commission for Research into European Food History (ICREFH) was founded in Münster, Germany, in 1989 by Professor Hans-Jürgen Teuteberg. The main objective of the Commission is to serve as a network for scholars interested in the issues of the history of food in Europe and to organize biennial symposia on relevant topics of research.
2. See Fowler, Peter J., *Then, now: the past in comtemporary society* (London 1995, 1st edn 1992), 41–42.

The previous volumes published by ICREFH are:

Teuteberg, Hans-Jürgen (ed.), *European Food History. A research review* (Leicester, 1992).

Burnett, John & Oddy, Derek J. (eds), *The Origins and Development of Food Policies in Europe* (Leicester, 1994).

Den Hartog, Adel P. (ed.), *Food Technology, Science and Marketing: European diet in the twentieth century* (East Linton, 1995).

Schärer, Martin R. & Fenton, Alexander (eds), *Food and Material Culture* (East Linton, 1998).

Fenton, Alexander (ed.), *Order and Disorder: The Health Implications of Eating and Drinking in the Nineteenth and Twentieth Centuries* (East Linton, 2000).

Part 1
Feeding the towns

PETER A. KOOLMEES

2 Veterinarians, abattoirs and the urban meat supply in the Netherlands, 1860–1940

During the last two decades, the academic world has increasingly turned its attention to studying the history of food, and its relationship to the health and disease of man and animals.[1] In these studies, much attention has been paid to the role of physicians and chemists in safeguarding food quality and hence in improving public health.[2] The attention paid to the veterinary contribution to food quality control has remained, however, rather limited. According to Rosser, the role of veterinarians has been somewhat underestimated.[3] Until recently, the history of veterinary public health[4] was recorded only by veterinarians. These studies dealt mainly with scientific and technical advances, the association between animal disease problems and public health, and the establishment of national and international organizations.[5] Little attention has been paid to the related socio-economic context, the historiography of abattoirs[6], and meat hygiene. Veterinarians have published a number of local studies on the history of meat inspection and abattoirs, while a comprehensive literature on slaughterhouse construction throughout the twentieth century has been produced by architects, particularly in France and Germany. Only recently have researchers from other disciplines focused their attention upon the development of meat hygiene and abattoirs and addressed these topics more from a societal point of view.[7]

During the process of transformation from agricultural and artisan societies to industrial and urban ones, fundamental changes in public health care, eating habits, foodstuff production, distribution and quality control occurred.[8] Inevitably, these changes affected the meat supply chain. The availability and supply of safe and sound meat became one of the problems which confronted local and national authorities in Europe and North America during this transformation. The authorities had to harmonize the economic interests of meat producers, meat traders and butchers with caring for the urban environment and the health of the citizens. Confronted with these conflicting interests, governments resorted to a bureaucratic solution, instituting mandatory centralized slaughtering in abattoirs under professional supervision.

In order to obtain a deeper insight into the emergence and development of abattoirs in the Netherlands from the mid-nineteenth century until 1940, the following questions will be considered in this chapter. Which actors played a significant role in the process of instituting the sanitary reform of the urban meat supply by developing public slaughterhouses and establishing mandatory meat inspection by expert veterinarians? In which way did the veterinary profession become involved in the quality control of foods of animal origin, particularly meat hygiene? And finally, what priority was given to abattoirs by the local authorities responsible for public health policy, as compared with other collective sanitary provisions instituted in the same period?

Meat production and consumption

From the middle of the nineteenth century onwards, north-west Europe underwent a gradual transition from a traditional rural society to a modern welfare state. This process of modernization was accompanied by characteristic changes including strong population growth, urbanization, economic growth, industrialization and a gradual rise in the standard of living. These changes caused a shift in food consumption patterns, resulting in, among other things, an increase in meat consumption. This was particularly the case in industrialized countries like Great Britain, France and Germany, where the situation of the working class improved. The British market was crucial for meat exporters. Due to extensive livestock farming, new transport possibilities and innovations in cooling and freezing technology, meat exports from sparsely populated countries (for instance, the USA and Argentina) to Europe, increased. Consequently, a large-scale international meat trade developed. As a result of increasing domestic demand for meat and ample opportunities for meat and meat product exports, cattle breeding and meat production in the Netherlands expanded from 1870 onwards.[9]

Around the middle of the nineteenth century, physicians began to criticize the lack of proteins of animal origin in the diet of the working class and poor, suggesting that it resulted in poor health, lack of physical strength and low productivity. A high level of meat consumption became the priority of not only food scientists, physicians and veterinarians, but also politicians, economists and the military. Improvements in both the quantity and quality of meat were a prerequisite to this attitude. Akin to the more industrialized neighbouring countries, a quantitative and qualitative improvement in diet occurred in the Netherlands and the democratization of meat consumption gradually began. In this sense, democratization means that meat was no longer exclusively a food for the rich, but that sufficient meat of various kinds was consumed by the majority of the population.[10] Between 1850 and 1930, the annual per capita meat consumption in the Netherlands increased from 27 to 50 kg. However, an assessment of the meat supply in the second half of the nineteenth century reveals that the soundness of meat, particularly cheap meat that was usually consumed by working class people, left much to be desired.

Meat supply and quality

Increased meat consumption necessitated a large-scale supply of sound meat. However, radical reforms in organization and hygiene were needed to accomplish this goal. Before the emergence of urban societies and their cattle and meat markets, consumers raised their own livestock and provided themselves with meat. In the countryside, home slaughtering remained common until well into the twentieth century when household refrigeration was introduced. In most cities, animals were slaughtered by butchers in small, privately owned butcheries. Beginning in the late Middle Ages, the slaughtering, selling and inspection of meat was regulated by butchers' guilds and local authorities. After the French Revolution, the guilds were abolished, and local meat inspection and meat trade regulations were disregarded. From around 1870 onwards, population growth and urbanization went hand in hand with an expanded network of these small butcheries, where slaughtering often took place under poor hygienic conditions. The city of Utrecht, for instance, with a population of about 100,000 in 1890, had 114 registered private butcheries, 378 'shops' where meat was sold, and a market for cheap meat and sausages. In addition, meat from knackers' yards was marketed or processed into pies and sausages and sold to labourers and the poor. Regrettably, the meat often originated from animals infected with anthrax, tuberculosis, trichinosis and tape-worms, as well as meat from animals that had died. A mere four inspectors were appointed to carry out meat inspection for the city of Utrecht.[11]

It is clear that, under these circumstances, consumers were left virtually unprotected from fraud and the adulteration of meat and meat products. The calamities that followed were almost inevitable. This situation was complicated by the increasing size, extent and complexity of the supply chain between producers and consumers. The poor hygienic conditions, which characterized meat and sausage processing, were due to a lack of hygiene awareness. Increasing numbers of butcheries represented a nuisance for the citizens. The transport and slaughtering of animals, as well as the storage and transport of offal within the city walls, led to a deterioration of the urban environment. Numerous outbreaks of trichinosis and meat poisonings occurred, infecting hundreds and killing dozens of people. Consequently, local and national authorities were increasingly confronted with complaints about the filth that butcheries caused in the urban centres and the poor quality of the meat offered.[12] Mass outbreaks of meat-borne diseases alarmed the authorities and clearly demonstrated the need for meat hygiene control. Politicians and local authorities were more or less forced to develop a public health policy related to the meat supply.

Health boards, veterinarians and meat hygiene

Improvement of the meat trade and meat inspection regulation, and of the urban environment became a regular issue in local and national politics around

1850. A programme was available which contained the two key factors for effective meat inspection: mandatory inspection by professional veterinarians and centralized slaughtering in public abattoirs located on the outskirts of towns. Such a programme had already been drawn up by French and German scientists such as Antoine Laurent Lavoisier (1743–1794) and Johann Peter Frank (1745–1821) at the end of the eighteenth century. One concept of the Enlightenment was that the state should assume responsibility for social and economic welfare. In the second half of the eighteenth century this idea was implemented, especially in France, Austria and Prussia, where the state intervened in agriculture as well as public health. The realization of public health programmes was promoted by local health boards, but above all required active state interference by means of a centralized bureaucracy before they became effective.[13] Following the French example, health boards were established in most larger towns in the Netherlands in the second half of the nineteenth century. Within these boards, the hygienists, a group of progressive physicians, engineers, physicists, chemists, lawyers and veterinarians, played a predominant role. Their great accomplishment was to turn public health into a political issue. They tried to realize this by means of professionalization and a scientific approach to public health care. Local and national health boards served as their forum. Due to their knowledge in the field of food hygiene, veterinarians participated in these public health boards from the outset.[14]

In the first half of the nineteenth century, there was essentially no scientifically-based meat inspection in Western Europe and hardly any veterinarians were involved in meat inspection. The untrained meat inspectors who were appointed in the larger towns relied almost exclusively on empirical knowledge. From 1850 onwards, veterinarians involved in health boards offered advise on all matters concerning veterinary public health due to their specific knowledge of zoonoses and their ability and training in recognizing diseased animals. They focussed on a scientifically and legally based form of meat inspection, research on meat poisoning, humane slaughtering methods, and regulations for the collection and destruction of waste from slaughterhouses and butcheries. Initially, it was mainly physicians who wrote about the subject of meat hygiene. Nevertheless, in various countries, a number of veterinarians started to publish on the subject, claiming this new field of professional activities for themselves. They stressed the importance of veterinary medicine for society as a whole, not only in the maintenance of a prosperous livestock industry, but also because of its value in safeguarding the healthy quality of foods of animal origin. The efforts of veterinarians to promote veterinary public health as a new field were strongly supported by the Netherlands Veterinary Association. This association recognized the potential for job opportunities in meat inspection services and public slaughterhouses and a possible broadening of the legal basis of the veterinary profession.[15] These claims, however, required a scientific basis of meat inspection.

The standing of meat inspection as a veterinary discipline increased when the scientific backgrounds of several diseases related to meat consumption were discovered one after the other. Firstly, there were the findings in the field of parasitology. The cycle of tapeworms was discovered in the 1860s and, by incision, meat inspectors were able to apply a preventive control. The occurrence of parasites declined, especially where trichinella control was instituted. From 1880 onwards, meat hygiene research obtained a more scientific character and bacteriological meat research was carried out in slaughterhouse laboratories. Between 1890 and 1940, a number of meat-borne pathogens were isolated.[16]

Around 1900, the traditional empirical meat inspection was transformed into an applied veterinary science. Research findings were published in handbooks and professional literature, and theoretical and practical meat inspection became part of the veterinary curriculum. National and international associations for professional veterinary food inspectors came into being. Meat hygiene became an important issue for international congresses on veterinary medicine and public health. All these factors contributed to the development of a professional meat inspection corps. Veterinarians had succeeded in providing an adequate answer to the mounting concerns for the more strictly public health aspects of the meat trade. The extension of their professional activities was laid down in meat hygiene legislation. As a result of the contribution of veterinarians in the discussions in municipal and national health boards, authorities reinforced or updated their regulations on meat inspection. A large number of towns made plans to establish public abattoirs. Often, it was veterinarians who initiated these plans and who drafted reports on the feasibility of these institutions. In addition to local activities, numbers of veterinarians were involved in formulating national meat inspection legislation. Between 1901 and 1922 a series of laws concerning meat inspection was adopted and enforced. Respectively these were: a revision of the Nuisance Act in 1900 facilitating municipalities to establish public abattoirs; an Act Regulating the Inspection of Meat for Exports in 1902 (revised in 1907 and 1920); and a Bill for Domestic Meat Inspection in 1913 which was enacted in 1919 and enforced in 1922. Within this framework lawyers, physicians and veterinarians participated in state committees with the task of drawing up meat inspection regulations that involved centralized slaughtering in public abattoirs with mandatory meat inspection under veterinary supervision based on uniform guidelines.[17]

Abattoirs and 'slaughter-warrant'

From 1795 onwards, attempts were made in a few Dutch towns to establish public slaughterhouses. These attempts, however, failed. As for slaughterhouses, Napoleonic France was the European leader in this area; until well into the nineteenth century, the Parisian abattoirs (built in 1810–1814) were regarded as exemplary. The butchers of Paris, and later of all larger French cities, were forced to slaughter their animals in the new public slaughterhouses, while the small, privately owned butcheries were all closed down. The French example was followed in other European countries. From around 1880, Germany, Switzerland and Austria-Hungary took the lead in building modern slaughterhouses from a technical, as well as a hygienic, point of view. England, Spain, the Netherlands and the Scandinavian countries lagged behind. Until the 1880s, only three municipal slaughterhouses were established in the Netherlands (Maastricht, Venlo and Bois-le-Duc).[18]

Generally, there was strong opposition to abattoirs from butchers in all European countries. Having played an essential role in the municipal meat supply for centuries, the butchers feared that the municipal slaughterhouses would limit their profession considerably. For one thing, they would be forced to slaughter their cattle at the public abattoir under supervision. The competency of municipal authorities to establish a public slaughterhouse and to prohibit private slaughtering, the so-called 'slaughter-warrant', became a politically contentious issue in the nineteenth century. In a number of countries, this matter was settled by passing a nuisance law. In the Netherlands, this warrant was instituted by the Nuisance Acts of 1875 and 1901.

Most local authorities hesitated in establishing municipal slaughterhouses due to the large financial expenditure required, possible increases in meat prices, and the question of profitability. Competition for local financing also came from other large and expensive projects in the field of sanitary reforms. In the second half of the nineteenth century, lengthy debates between supporters and opponents of municipal abattoirs took place within municipal councils. Often, several decades elapsed between the first proposal for establishing a municipal slaughterhouse and the actual opening of the institution (see Table 2.1). Two factions could be distinguished in the councils: one side included hygienists, with their demand for adequate meat inspection, animal protectionists and citizens having nuisance complaints about the private butcheries; the other side was composed of a coalition of butchers and meat traders, whose independence was threatened by the establishment of municipal abattoirs. At first, economic interests and the objections of the butchers outweighed the arguments concerning public health and pollution put forward by the hygienists. By the turn of the century, the growth of socialism tipped the balance. Ultimately, though, the decision to establish a public slaughterhouse mostly depended on the financial position of the municipality and hence, on general economic fluctuations.[19]

Table 2.1 *Municipal slaughterhouses established in 15 Dutch towns in the period 1883–1910.*

Town	Proposal[1)]	Opening	Inhabitants	Cost (guilders)	Animals slaughtered [2)]
Rotterdam	1866	1883	162,000	800,000	38,000
Amsterdam	1856	1887	390,000	2,200,000	91,000
Roermond	1853	1899	13,000	110,000	12,000
Elburg	1898	1900	2,700	2,000	n.a.
Groningen	1872	1900	67,000	430,000	36,000
Nijmegen	1875	1900	46,000	260,000	15,000
Maastricht	1876	1901	34,000	305,000	26,000
Utrecht	1866	1901	102,000	570,000	26,000
Leiden	1895	1903	55,000	328,000	16,000
Dordrecht	1894	1906	44,000	175,000	9,000
Haarlem	1878	1907	70,000	416,000	19,000
Alkmaar	1891	1908	21,000	250,000	10,000
Sittard	1892	1908	7,000	47,000	5,000
Arnhem	1902	1910	65,000	640,000	22,000
Bois-le-Duc	1898	1910	35,000	183,000	11,000

Source: Koolmees (1997), 150, 170.

Note: [1)] First proposal in city council.

[2)] Number of animals slaughtered in the first year after opening.

The passing of the Dutch Meat Inspection Act in 1919 stimulated the building of public slaughterhouses indirectly, since this Act imposed hygienic requirements concerning furnishing and equipment of private abattoirs, butcher's shops etc., and outlined the obligation of local authorities to establish a meat inspection service. Many existing privately-owned butcheries could not meet these new requirements and were forced to close, prompting the authorities of many cities to build public slaughterhouses in order to institute an adequate meat inspection service. Between 1883 and 1940, the Netherlands were covered by a network of 86 municipal slaughterhouses (see Figure 2.1). Apart from the positive influence of the Meat Inspection Act and the profitability of the existing abattoirs, it was mainly the favourable financial position of the cities in the period 1922–1929 that contributed to this rapid spread of municipal slaughterhouses. In 1927, not fewer than 15 abattoirs became operative.

According to the Dutch Meat Inspection Act, only graduated veterinarians or trained lay-inspectors under the supervision of veterinarians were allowed to conduct meat inspection. The municipalities were responsible for the implementation of the Act. The country was divided into 402 different inspection districts, each headed by a veterinarian. This function was often combined with that of director of the public slaughterhouse when that facility was present in the same district. An increasing proportion of veterinarians found employment in slaughterhouses and meat inspection services. As a result of centralized slaughtering under professional supervision, meat inspection became much more effective. The number of infections with parasites was effectively diminished, tuberculosis was better detected and the prevalence of home slaughtering decreased.[20]

Figure 2.1 *Establishment of municipal slaughterhouses in the Netherlands in the period 1883–1940: total number (left y-axis) and number per year (right y-axis).*

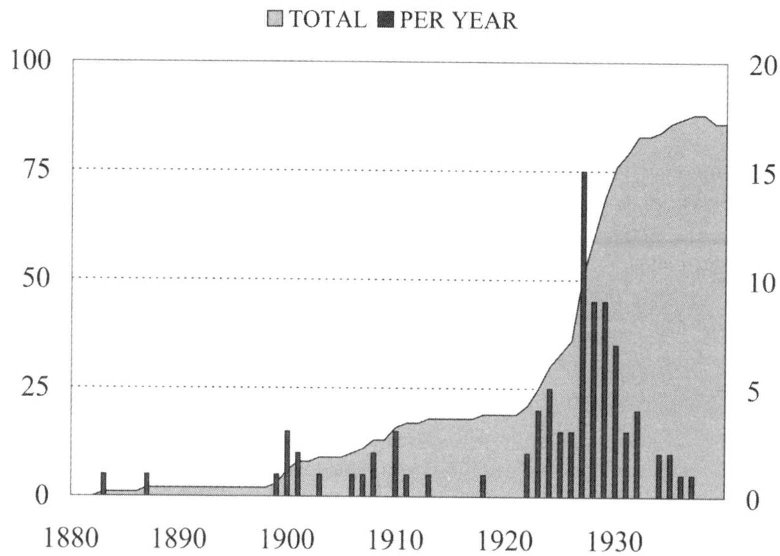

Source: Koolmees (1991), 134–135.

Abattoirs and other sanitary provisions

A number of municipalities did not wait for the establishment of a national meat inspection regulation but proceeded to institute a public slaughterhouse on their own. In 1922, the year in which the national Meat Inspection Act became operative, there were already 21 public abattoirs in the Netherlands. Food inspection services followed a similar trend. A Food Inspection Act was discussed in parliament from the turn of the century onwards, but became operative only in 1921. At that time, not fewer than 21 food inspection services were active.[21]

The schedule upon which municipal slaughterhouses were established varied considerably from town to town. In this respect, it is interesting to examine the implementation pattern of the establishment of collective sanitary provisions. In Table 2.2 a comparison of the implementation schedule of municipal slaughterhouses and four other sanitary provisions in a number of towns is presented.

Table 2.2 *Establishment of five different sanitary provisions in 20 Dutch towns.*

Town	Waste disposal (barrel-system)	Water-works	Public cleaning	Food inspection	Public abattoirs
Amersfoort	1883	1890	1884	1913	1927
Amsterdam	1876	1853	1877	1858	1887
Arnhem	1871	1885	1874	1911	1910
Deventer	1875	1893	1875	1913	1929
Dordrecht	1874	1882	1918	1907	1906
Enschede	– –	1892	1898	1921	1922
The Hague	1871	1874	1871	1907	1911
Groningen	1700s	1881	1855	1908	1900
Haarlem	1877	1898	1877	1911	1911
Den Helder	– –	1854	1885	1913	1965
Bois-le-Duc	1874	1887	1899	1913	1910
Hilversum	– –	1886	n.a.	1913	1924
Leeuwarden	1872	1888	1870	1916	1925
Leiden	– –	1878	1914	1901	1903
Maastricht	– –	1887	1884	1911	1901
Nijmegen	1877	1897	1877	1906	1901
Rotterdam	1877	1874	1876	1893	1883
Tilburg	– –	1898	1903	1913	1924
Utrecht	1876	1883	1876	1913	1901
Zutphen	1875	1889	n.a.	1921	1930

Sources: Halbertsma (1899), 161–162; Koolmees (1991), 134–135; Van Raalte (1928), 369; De Zeeuw (1983), 17; Van Zon (1986), 85, 263.

At first sight, there hardly seems to be a distinct pattern. The period in which these provisions were instituted varied considerably between towns and, within a town, between the different provisions. Only roughly can a more detailed distinction between 'early' and 'late' towns be made. The larger towns in the western part of the country were generally early, but the same goes for Groningen and Nijmegen. Deventer, Enschede and Tilburg can be considered late adapters. However, the most striking point is the large difference in pace with which the different provisions in one town were established. For example, The Hague was quite early with the institution of waterworks (1874), but for a large town, very late in building a slaughterhouse (1911). This can also be observed in Den Helder where these two provisions were instituted in 1854 and 1965, respectively.

On average, waste disposal, waterworks and public cleaning departments came into being earlier than food inspection services and abattoirs. Waterworks were constructed at an early stage in Amsterdam and Den Helder, the other towns followed in the 1880s. Most municipal cleaning departments were established in the 1870s. Food inspection services were instituted in the first two decades of the twentieth century, with a distinct peak in 1913. The majority of municipal slaughterhouses came into being in the 1920s. Hence, it seems that waterworks, waste disposal and public cleaning departments were considered among the most urgent sanitary provisions. With some caution, it can be concluded that, compared with the other provisions mentioned, a lower priority was given to public slaughterhouses. Indeed, these institutions were considered more desirable than necessary (in terms of public hygiene) by the local authorities.[22]

Of course, it was likely that the expenditure required to establish these sanitary provisions played a crucial role in determining their necessity. Regarding waste disposal, for example, at first most towns chose to introduce the relatively cheap barrel-system instead of the installation of an expensive large-scale sewer system. In the period 1900–1910, the expenditure for a municipal food inspection service amounted to a few thousand guilders. As shown in Table 2.1, the installation costs of abattoirs were much higher. Therefore, public slaughterhouses were built when municipalities had sufficient financial means. The financial position of municipalities was determined by the income from both local taxes and state subsidy. Consequently, the establishment of abattoirs was strongly dependent on general economic fluctuations. Apart from these financial considerations, the protracted debates between supporters and opponents of abattoirs in local politics mentioned above contributed to delay in the establishment of public slaughterhouses.

Conclusion

Around 1900, the Netherlands bore witness to a gradual change from decentralized slaughtering and marginal inspection in numerous small butcheries to the centralization of these activities in public slaughterhouses. In initiating and guiding this fundamental change, a group of hygienists that was active in health boards played a predominant role. Veterinarians were represented in this group. They contributed significantly to the scientific development of meat inspection. As a new domain of professional activities, municipal slaughterhouses and meat inspection played an important role in the professionalization process of the veterinary profession. Under the Meat Inspection Act, responsibility for a large segment of public health was entrusted to this profession. Between 1922 and 1940, the supply of sound meat was improved considerably by the network of municipal slaughterhouses and meat inspection legislation.

Abattoirs and regular meat inspection can be considered collective sanitary provisions that were established within the framework of municipal health care. Based on the liberal ideology of free trade and self-help, state interference in the meat supply chain remained limited in the nineteenth century. The spread of hygienic awareness led to a change in the attitude of local and central authorities who gradually took more trouble to improve public hygiene. Compared with other sanitary provisions, however, a lower priority was given to abattoirs.

NOTES

1. Mennell et al. (1992); Teuteberg (1992); Burnett & Oddy (1994).
2. Hanssen & Wendt (1965); de Zeeuw (1983); Hutt & Hutt (1984); Ellerbrock (1987); Jones (1992); Burnett & Oddy (1994).
3. Rosser (1991).
4. Defined as all interactions between animals and animal products on the one side and human health on the other.
5. Schwabe (1984); Grossklaus et al. (1991); Reuter (1997).
6. Also referred to as public abattoirs, public or municipal slaughterhouses.
7. Burgholtz (1987); Ponthieux & Philipp (1987); Vialles (1987); Wade (1987); Koolmees (1991, 1997); Koolmees et al. (1999).
8. Mennell et al. (1992), 68–74; Porter (1994).
9. Perren (1978); Walsh (1982); Diederiks et al. (1994).
10. den Hartog (1987), 24; Koolmees (1997), 45–53, 90–92.
11. Koolmees (1997), 109–112.
12. van Daalen (1986), 280–282; Koolmees et al. (1999).
13. Moreau (1916), 3–6; Fowler la Berge (1975), 339–340.
14. Houwaart (1991); 19–21; Koolmees (1997), 75–82, 109–110.
15. von Reeken (1861), 46–56; Sticker (1890–91); Morris (1895), 78.
16. Koolmees (1997), 70–75, 99–105, 122–129.
17. Schwabe (1984); Koolmees et al. (1999).
18. de Loverdo (1906), 691–859; Heiss (1912), 9–25.
19. Koolmees (1997), 145–152.
20. *Annual Reports of the State Veterinary Service* (1920–1940).
21. van Eekelen (1984), 211.
22. Koolmees (1997), 272.

BIBLIOGRAPHY

Annual Reports of the State Veterinary Service (The Hague, 1920–1940).

Burgholtz, Dieter, 'Die wirtschaftliche Entwicklung von Märkten, Messen und Schlachthöfen (ab ca. 1850 bis zur Gegenwart)'. *Zeitschrift für Unternehmensgeschichte*, Beiheft 42 (1987), 88–124.

Burnett, John & Oddy, Derek J. (eds), *The Origins and Development of Food Policies in Europe* (London / New York, 1994).

Daalen, R. van, 'Nauwbehuisd en dichtbevolkt. Gemeentelijke voorzieningen en klachten uit de burgerij, Amsterdam 1865–1920' [Cramped and densely populated. Municipal provisions and complaints from citizens, Amsterdam 1865–1920]. *Sociologisch Tijdschrift* 12 (1986), 274–307.

Diederiks, H. A. et al., *Van agrarische samenleving naar verzorgingsstaat. De modernisering van West-Europa sinds de vijftiende eeuw* [From agricultural society to welfare state. The modernisation of Western Europe since the fifteenth century], 2nd edn (Groningen, 1994).

Eekelen, A. van, *Naar een rationele zuigelingenvoeding. Voedingsleer en kindergeneeskunde in Nederland 1840–1914* [Towards a rational infant feeding. The science of nutrition and paediatrics in the Netherlands 1840–1914] (University of Nijmegen Ph.D., 1984).

Ellerbrock, Karl-Peter, 'Lebensmittelqualität vor dem Ersten Weltkrieg: Industrielle Produktion und staatliche Gesundheitspolitik', in: H. J. Teuteberg (ed.), *Durchbruch zum modernen Massenkonsum. Lebensmittelmärkte und Lebensmittelqualität im Städtewachstum des Industriezeitalters* (Münster, 1987), 128–188.

Fowler la Berge, A., 'The Paris health council, 1802–1848'. *Bulletin of the History of Medicine* 49 (1975), 339–352.

Grossklaus, D. et al., 'Notes on technical progress in veterinary public health'. *Revue Scientific et Technique Office International des Épizooties* 10 (1991), 995–1018.

Halbertsma, H. P. M., 'Historische schets der drinkwaterleidingen in Nederland' [Historical review of waterworks in the Netherlands], in: *De ziekenverpleging en de zorg voor de openbare gezondheid in de laatste 50 jaren, Issue Nederlandsch Tijdschrift voor Geneeskunde* (Amsterdam, 1899), 158–163.

Hanssen, E. & Wendt, W., 'Geschichte der Lebensmittelwissenschaft', in: J. Schormüller (ed.), *Handbuch der Lebensmittelchemie*, vol. 1 (Berlin, 1965) 1–75.

Hartog, A. P. den, 'Voedingsgewoonten en een veranderende samenleving' [Food habits and a changing society], in: A. de Knecht-van Eekelen & M. Stasse-Wolthuis (eds), *Voeding in onze samenleving in cultuur-historisch perspectief* (Alphen aan den Rijn / Brussel, 1987), 15–37.

Heiss, H. A., *Bau, Einrichtung und Betrieb öffentlicher Schlacht- und Viehhöfe*, 4th edn (Berlin, 1912).

Houwaart, E. S., *De hygiënisten. Artsen, staat & volksgezondheid in Nederland 1840–1890* [The hygienists. Physicians, state and public health in the Netherlands 1840–1890] (Groningen, 1991).

Hutt, P. B. & Hutt, P. B. II, 'A history of government regulation of adulteration and misbranding of food'. *Food, Drug and Cosmetic Journal* 39 (1984), 2–73.

Jones, J. M., *Food safety* (Minnesota, 1992).

Koolmees, P. A., *Vleeskeuring en openbare slachthuizen in Nederland 1875–1985* [Meat inspection and public slaughterhouses in the Netherlands 1875–1985] (Utrecht, 1991).

Koolmees, P. A., *Symbolen van openbare hygiëne. Gemeentelijke slachthuizen in Nederland, 1795–1940* [Symbols of public hygiene. Municipal slaughterhouses in the Netherlands, 1795–1940], University of Utrecht Ph.D. (Rotterdam, 1997).

Koolmees, P. A. & Fisher, J. R. & Perren, R., 'The traditional responsibility of veterinarians in meat production and meat inspection', in: F. J. M. Smulders (ed.), *Veterinary aspects of meat production, processing and inspection. An update of recent developments in Europe* (Utrecht, 1999), 7–30.

Loverdo, J. de, *Les abattoirs publics, Vol. 1: Construction et agencement des abattoirs* (Paris, 1906).

Mennell, Stephen & Murcott, Anne & Otterloo, Anneke H. van, *The Sociology of Food: Eating, Diet and Culture* (London / New Delhi, 1992).

Moreau, A., *l'Abattoir moderne, construction, installation, administration*, 2nd edn (Paris, 1916).

Morris, C. D., 'Meat inspection'. *Journal of Comparative Medicine and Veterinary Archives* 16 (1895), 71–78.

Perren, R. *The meat trade in Britain 1840–1914* (London, 1978).

Ponthieu, Gérard & Philipp, Elisabeth, *La Vilette, les années 30. Un certain âge d'or* (Paris, 1987).

Porter, D., *The history of public health and the modern state* (Amsterdam, 1994).

Raalte, A. van, 'De ontwikkeling van het keuringsdienstwezen in Nederland' [The development of food inspection in the Netherlands]. *Chemisch Weekblad* (1928), 368–370.

Reeken, C. G. von, 'De geneeskunde der dieren in verband met staats- en volksbelang' [Veterinary medicine and its relation to state and public interest]. *De Gids* 25 (II) (1861), 27–56.

Reuter, G., 'Veterinärmedizin und Gesundheitsvorsorge'. *Berliner/Münchener tierärztliche*

Wochenschrift 110 (1997), 431–435.

Rosser, W. W., 'Corollary development of the professions of veterinary medicine and human medicine in the United States'. *Veterinary Heritage* 14 (1991), 3–29.

Schwabe, C. W., *Veterinary medicine and public health*, 3rd edn (Baltimore, 1984).

Sticker, A., 'Zugehörigkeit der Fleischbeschau zur thierärztlichen Thätigkeit'. *Archiv der animalische Nahrungsmittelkunde* 6 (1890–1891), 109–112, 121–123.

Teuteberg, Hans J. (ed.), *European Food History. A Research Review* (Leicester / London / New York, 1992).

Vialles, Noélie, *Le sang et la chair. Les abattoirs des pays de l'Adour* (Paris, 1987).

Wade, L. C., *Chicago's pride. The stockyards, packingtown, and environs in the nineteenth century* (Urbana / Chicago, 1987).

Walsh, M., 'From pork merchant to meat packer: the midwestern meat industry in the mid nineteenth century'. *Agricultural History* 56 (1982), 127–137.

Zeeuw, P. W. de, *Gewogen en te licht bevonden. 75 jaar Keuringsdienst Dordrecht* [Tried and found wanting. Commemorating the 75th anniversary of the Food Inspection Service Dordrecht] (Dordrecht, 1983).

Zon, H. van, *Een zeer onfrisse geschiedenis. Studies over niet-industriële vervuiling in Nederland, 1850–1920* [A very shady history. Studies on non-industrial pollution in the Netherlands, 1850–1920] (Groningen, 1986).

HANS-JÜRGEN TEUTEBERG

3 Cultivation, markets and the rise of vegetable consumption in German towns, 1850–1913

The explosive urban growth which began in some European areas during the nineteenth century is perceived, like the transition to industrial work, as an autonomous determinant of modernization. Both processes were accompanied by an almost totally new form of urban food supply. This last mentioned far reaching structural change has been focussed on more and more by historians in recent decades.[1] It is understandable that the growing urban demand for the basic victuals – bread, meat and potatoes, was first to be analyzed. From studying consumption per head, from family budgets, advertisements and cookery books, it is obvious that the consumption of fruit and vegetables rose in the late nineteenth century probably more than ever before or afterwards. This signifies an enormous improvement in the quality of common food.[2] Since cultivation, trade and consumption of fruit in Germany has been studied recently, green vegetables should now be examined in the same manner in order to allow comparisons with other European countries.[3] Our investigation in this chapter centres around the question of what the growing consumption of vegetables had to do with urbanization, and furthermore with the more longitudinal influences of improvements and rationalization processes in the fields of production, transport and trade. But before we turn to this rise in the consumption of vegetables we have to define this food product precisely, because for a non-botanist it is not always clear.[4]

The German collective term *Gemüse* (vegetables) can be traced back to the old high-German word *muos*, which meant originally the whole fare of pap. This seems to hint that for many centuries all edible plants were consumed only when cooked. The expression *gemuese* is then found in the middle high-German language. Today this term covers all parts of plants ripening within one year and which are served fresh and raw, cooked, preserved or otherwise prepared – leaves, flower-buds, roots, tubers, bulbs, parts of stalks, sprouts, blossoms, fruits and seeds. The fruits of plants growing several years are counted as fruit. Furthermore, we can distinguish the following main species of vegetables:

- *Salad vegetables:* lettuce, endive, spinach, chard (silver beet), lamb's lettuce, chicory, parsley;

- *Fruit vegetables:* several sorts of beans and peas, tomatoes, paprika (red pepper), cucumber, pumpkins, water and sugar melons;
- *Root vegetables:* carrots, radishes, white and red beet, turnips (Swedes), comfreys, horse-radish;
- *Cabbage vegetables*: cauliflower, broccoli, white and red cabbage, savoy cabbage, chinese-cabbage, kohlrabi, Brussels sprouts, kale (borecole);
- *Bulb vegetables:* onions, garlic, leek and chives;
- *Stalk vegetables and mushrooms:* asparagus, rhubarb, mushrooms, several sorts of celery;
- *Other not so well known vegetables:* egg plants, artichokes, garden cress, fennel, corn-on-the-cob.

All these kinds of green vegetables we can list further after their botanic families. The pulses (beans, peas and lentils) which are stored in a dried form and then cooked later, are not counted as vegetables nor are potatoes.

Kitchen gardens and weekly markets before the beginning of modern urbanization

Many vegetable plants in Germany were generally named in former times *Kraut* or *Grünzeug* having the same meaning as in English greens or in Dutch *kruit*. The German expression was originally a collective term for all small plants, later only for their leaves (today the forms *Kartoffelkraut* [potato leaves], *Rübenkraut* [root leaves] and *ins Kraut schiessen*, i.e. quick growing plants, are archaic). Later only those parts of plants called *Kraut* were used for food, remedies or spices. It is worth mentioning that plants of no use for gardening or agriculture are today named *Unkraut* (weeds), which has a negative meaning. These sort of plants have therefore to be destroyed. At least *Kraut* means only special sorts of useful vegetables. The greengrocer (*Gemüsehändler*) is a quite modern expression. Until the end of the nineteenth century they were often named *Kräutler* and sold vegetables at the *Krautmarkt* in older small provincial towns. The name of the famous vegetable dish *Sauerkraut* is also a reflection of this old meaning. But we must also remember *Bohnenkraut* (herb beans), *Kräutertee* (herb tea) and *Kräuterkäse* (herb cheese). Finally, a man is called in German a *Krauter* when he is ill. In former times, if somebody was dying and all medical treatment was in vain, people used to say: "*Dagegen war kein Kraut gewachsen*" (There was no kraut against this). As these examples demonstrate common vegetables were considered as in antiquity as food, spice and remedy together.

Up to the eighteenth century and the beginning of the modern natural sciences, folk medicine held fast to the traditional superstition that each illness had its suitable plant for healing purposes. Therefore after the invention of printing several *Kräuterbücher* or *Herbarien* (herbals) were published from the sixteenth century onwards.[5] They show that besides the small kitchen garden many other wild plants were regularly collected like sorrel, stinging nettle, dandelion, wild chicory, cress and scurvy-grass. Apart from pulses only some sorts of cabbage, roots, pumpkins and cucumber were suitable as winter food. These vegetables were pickled with salt, vinegar or mustard and flavoured with juniper-berries, dill caraway-seeds (cumin). The head of the white cabbage was cut with a special knife into small pieces and put together with salt layers into a wooden barrel. With the help of a cabbage press (*Krautpresse*) or the weight of a heavy stone, a lactic acid with a high content of vitamin C was produced which preserved cabbage for four months. This product was named in Germany *kumpost* which comes from the Latin *compositas*. Later on the expression *sûrkrût* or sauerkraut was used. Due to its freshness, fine sour taste and healthy quality, this pickled white cabbage ranked, at the top of the winter vegetables. In many kitchens in Middle Europe the so-called cabbage cask (*Krautfaß*) was a necessary utensil. Other sorts of cabbage and root were stored in the cellar or protected with earth against winter frost, but never enjoyed the widespread popularity of sauerkraut.

Citizens in the late Middle Ages and early modern times often had the opportunity to cultivate vegetables in their own or rented kitchen gardens inside or outside of the town walls. Because this small private production was not sufficient, additional weekly markets were always necessary. Peasant women from the nearby surrounding countryside sold foodstuffs there. Due to the very high water content and the danger of rottenness long distance trade with vegetables was not possible. Old town records show that the women hawkers (*Hökerin*) had to pay a special fee (*Hockengeld*) to the market supervisor (*Hockenvogt*). The supervisor had besides other things to see that the mushrooms on offer were checked in case they were poisonous. Since the general knowledge about mushrooms was low, this foodstuff was scorned and detested in most German regions. By contrast with the Germans, their Slavonian and Latin neighbours evidently became mushroom eaters much earlier, but there is no clear reason why. The strongly held distrust of the Germans against mushrooms has probably to do with the fact that the fear of poisoning was always dominant rather than taste or nutritive value. In folk belief dead and mushrooms were seen as identical and therefore we can consider mushrooms as a sort of an anti-cultural foodstuff.[6] In reality there are approximately 150 kinds of mushrooms of which only a dozen are dangerous. This ratio is not much different from other herbs and berries which are well accepted. It seems remarkable that at the very end of the eighteenth century champignons and truffles imported from France appear on German menus for the first time as "curiosities". During the nineteenth century some recipes with mushrooms are mentioned in middle-class cookery books, but these were more exceptions for the tiny group of bourgeois gourmets. Around 1900 the large vegetable market near Hamburg's harbour offered beside imported dried truffles, morels and champignons, only fresh chanterelles (*Pfifferlinge*) and yellow boletus (*Steinpilze*) in small quantities at high prices. The lower social strata, especially country-families, remained strictly against mushrooms for much longer. There was still little knowledge about the species as the people's language shows. The expressions *Düwelspankoken* (devil's pancake) and *Taterfleesch* (gipsy meat) in Northern

Germany indicate until the early twentieth century deep religious and ethnic taboos against this sort of vegetable. The search for food substitutes during the First World War led to informative pamphlets and charts showing edible mushrooms and special urban testing laboratories, as well as the manufacturing of tinned mushrooms. More recently, East-German refugees and those expelled from the German-Slavonic border areas between 1945 and 1947 have caused a modest rise of consumption.

The transition to larger vegetable tillage caused by the growing urban demand in the late nineteenth century

Vegetable cultivation was intensified through the rising wave of rationalization in the agrarian economy during the seventeenth and eighteenth centuries.[7] Apart from the older more medical and botanical oriented herbals written by sixteenth-century humanists, a new 'husbandry-literature' was published which spread through many editions and copies all over Europe.[8] In these more practical instructions for the "Philosophical Peasant", "Thinking Economist" and the "Gentleman Farmer" all important vegetables plants were not only precisely described with engravings, but also experts gave detailed advice for cultivation, maintenance and control of parasites. Leaving aside all those plants which are primarily spices or remedies, then around 1800 we can count more than 30 species of vegetables at a minimum in Central Europe.[9] However, as this first increase in vegetable production was still restricted to small kitchen gardens as the titles of the books demonstrate, it is impossible to quantify its scale.

In the course of this initial process of agrarian rationalization, which was carried forward by numerous small and local Economic Societies (*Ökonomischen Sozietäten*) and later by larger provincial Associations of Garden Cultivation (*Gartenbau-Vereine*) the number of useful vegetable plants grew constantly, especially from the late eighteenth century onwards. So, in 1860, for example, more than 100 sorts of peas were known in Germany. In all these societies and their journals enlightened landowners discussed in detail the best varieties to respond to soil, climate variations, fertilizers and natural enemies of the plant, in order to get an optimal output, especially for the early ripening kinds of vegetables. The gardeners and farmers realized more and more that vegetable cultivation was the most profitable use of the soil and its productivity exceeded the traditional corn economy by far. Some regions in Germany, especially those near larger towns, gained a new prosperity, formerly unknown, by growing and selling vegetables and herbs. This new production was only exceeded in a few places by the cultivation of hops and tobacco. Nevertheless, it is true to say that up to 1860 all vegetable cultivation remained limited by their seasons. Fresh greens could mostly only be sold in spring and summer, because hot-houses heated by steam were too expensive for an ordinary vegetable farmer before the end of the nineteenth century.[10]

The Prussian statistician Georg von Viebahn reported after an inquiry into horticulture in the neighbourhood of big cities at this time that greens served to mark the differences between the social classes: "The so-called distinguished [people] are very dainty [...] they like a certain vegetable only so long as the poor can not have it, but they disdain it, if everybody can afford it; in any case they will it have earlier in the year."[11] Vegetables were used as some other foods to underline "conspicious consumption" (Thorstein Veblen) in respect of "fine social distinctions" (Pierre Bordieu). Here we touch an important point: until the beginning of rapid urbanization and the period of high industrialization the first fine greens in springtime were much more expensive than the normal coarse vegetables and especially the leguminous fruits. Fresh vegetables with a high water content up to 90 per cent (asparagus and tomatoes with even 99 per cent) were only good for very short distance transport and storage. With few exceptions greens had to be sold quickly, in order to avoid loss of quality. Up to the end of the First World War it was not known that fresh vegetables contained important vitamins, minerals and substances good for the digestition which got lost in cooking. For ordinary housewives with families on a small income, preparing a daily meal was the decisive factor. By contrast to the traditional pulses, greens demanded other foodstuffs in addition which could often be more expensive than the vegetables. Following Viebahn's first analysis which he made on the orders of the German Customs Union between 1858–1864, most Germans still consumed, as in former centuries, several sorts of roots and cabbage, carrots, cucumber and onions which could all be used for one-dish-meals. Many kinds of vegetables which we prefer today were not on the family table at that time.

Viebahn's first empirical survey about vegetable consumption in the whole of Germany makes clear that there existed very different regional food habits: The Bavarians, for instance, liked sauerkraut, radish, white and red beetroot but disliked carrots and green beans. The North Germans esteemed in contrast kale (*Grünkohl*) which was resistant to the frosts of winter. In Württemberg, down in the southwest, people still held fast to the mighty satisfying horse beans, a traditional hearty meal for hard working countrymen. Viebahn registered the general antipathy of the most Germans against spinach. Perhaps this had to do with its pappy dark green appearance and the fact that this cooked vegetable went bad very quickly. It is interesting that still today many housewives show a clear dislike against spinach without knowing why.[12] On the other hand the common preference for sauerkraut which Viebahn observed in all regions, caused foreigners often to believe that this was the real national dish of the Germans. In the United States, where more than five million Germans emigrated during the nineteenth century sauerkraut remained the symbol of cultural identity. Because most German immigrants stuck to their home recipes for at least one generation they got the nickname The Krauts from other Americans.

With the help of agrarian statistics between 1850 and 1913 we can demonstrate how the production of vegetables in the small kitchen gardens must presumably have decreased and on the other hand field cultivation increased, as Table 3.1 shows.

Table 3.1 *Vegetable crops in Germany in 1000 tons (t) 1850–1913.*

Year	(t)	Year	(t)	Year	(t)
1850	1,310	1871	2,050	1892	2,940
1851	1,340	1872	2,090	1893	2,990
1852	1,370	1873	2,140	1894	3,040
1853	1,410	1874	2,170	1895	3,080
1854	1,440	1875	2,220	1896	3,130
1855	1,480	1876	2,260	1897	3,180
1856	1,510	1877	2,290	1898	3,230
1857	1,540	1878	2,330	1899	3,280
1858	1,570	1879	2,370	1900	3,330
1859	1,610	1880	2,420	1901	3,380
1860	1,650	1881	2,460	1902	3,430
1861	1,680	1882	2,510	1903	3,480
1862	1,710	1883	2,550	1904	3,530
1863	1,750	1884	2,590	1905	3,580
1864	1,790	1885	2,630	1906	3,630
1865	1,830	1886	2,670	1907	3,680
1866	1,860	1887	2,720	1908	3,730
1867	1,900	1888	2,760	1909	3,790
1868	1,930	1889	2,800	1910	3,840
1869	1,970	1890	2,860	1911	3,890
1870	2,010	1891	2,900	1912	3,940
				1913	4,000

Source: Hoffmann (1965), 285–287 (table 50).

These figures document that vegetable production in the territory of the later German Reich more than tripled in 64 years and growth remained always constant.[13] However, greatly fluctuating crops meant much more uncertain incomes for farmers. On the other hand pulses lost their importance for daily fare by a comparable extent. Between 1805 and 1815 the areage in Prussia under legumes was three times larger than the area of potato cultivation. In 1862 this relationship had totally changed to the contrary. This rapid decrease of pulses continued; the consumption per head of pulses diminished from 20.7 kg in 1850 to 5.2 kg in 1913 and finally to 0.9 kg in 1975.

The supply of large towns with new forms of greengrocery during the late nineteenth century

The increasing need of large cities for agrarian produce and the growth of production set municipal administrations and food traders the task of organizing a totally new supply system. Until the end of the high-liberal period in Germany in 1880, the local magistracies controlled food provisioning in conjunction with the transport and customs policies of the provincial and national authorities. However, they abstained from all interventions in agrarian production and the food trade. Before the liberal reform era around 1800, urban administrations had always regulated the prices of victuals, their minimum weights and controlled food quality through their own commissioners. In times of food crisis these officials could enter houses by force, for example, to prevent the hoarding of flour. The liberal market regulations in the early nineteenth century abolished all these market interventions from the Ancien Régime. The new view was that the state, and especially local authorities, should not compete with the private economy. Urban communities should only intervene when food prices were monopolized by groups of farmers or tradesmen or when, after a bad harvest, urban supplies were in serious danger.[14] Such a crisis occurred in 1911, when the prices of potatoes and vegetables climbed to previously unknown high levels and influenced even the meat prices. Municipal sales offices were established in 41 German towns in order to protect households with small incomes. Through this massive intervention market prices fell shortly afterwards, when these government retail shops were immediately closed.[15]

A longer term task for the large towns was the building and operating of market-halls in order to improve traffic connexions and standards of food hygiene. The impulse for this new intervention was the Act in Respect to the Transport of Food and Luxury Foodstuffs and Other Objects of Use from the 14th of July 1879. This authorized the police to control all provision-dealers and to take samples of their goods for the newly erected official foodstuff laboratories, because the sale of adulterated or false victuals was strictly forbidden. In the case of danger to the consumer's health the law-courts could even impose prison sentences.[16] The new market-halls which were built in 20 German large towns before the end of the century offered many advantages especially for the trade in fruit and vegetables. They faciliated much quicker transport of foodstuffs which was especially important for all those goods with a high water content which were highly perishable. This had been always a major problem of the older weekly markets. Some of these modern market-halls had even their own railway-stations or harbour quays with equipment for mechanical unloading. The closed halls kept away heat, rain, cold and even dust, were centrally heated, illuminated and ventilated. They had their own water supply and cleaning service; moreover artificial refrigeration techniques made the preservation of highly perishable food possible. All these innovations had the economic effect of balancing the ups and downs of market prices and improving the food quality. Unsold stocks, which were very large in respect of fresh vegetables, could now be sold the next day without great losses and new transport costs. The market-halls for the wholesale and retailing trade corresponded, of course, with the newest technical standards. The buildings were constructions of steel and glass, with large cellars and galleries.[17] The cooling temperatures of fresh vegetables and the gains in keeping periods were, of course, very different, as Table 3.2. demonstrates.

Table 3.2 *Cold storage of vegetables in urban market-halls.*

Vegetable	Temperature (°C)	Duration
Beans	+3–4	10 days
Cauliflower	+0.5	8–10 weeks
Cabbage (white and red)	+0.5	9 months
Peas	0	4–7 weeks
Cucumber	0	2–4 weeks
Carrots	0	6 months
Brussels sprouts	–2.5	12 weeks
Salad (lettuce)	0	3–4 weeks
Celery	0	5–8 months
Asparagus	0	2–4 weeks
Spinach	0	6 weeks
Tomatoes	0	4–8 weeks
Onions	–2.5	7 months
Parsley, dill	–1.5	2 months
Rhubarb	0	2 months

Source: Ziegelmayer (1942), 140.

This table indicates that artificial cold storage in the market-halls with its graduated length of preservation represented a revolution for the whole vegetable trade. The extremely large loss of approximately 50 per cent of these highly perishable goods was considerably reduced. Due to this careful supervision of keeping times not only taste, appearance and nutritive value of vegetables, but also the general popularity of fresh vegetables among the consumers was improved.

The improvement of the transport system during the second half of the nineteenth century further promoted better quality vegetables. By its increased density after 1840, the new railway network in Germany contributed greatly to the provisioning of large towns. Highly perishable fruit and vegetables could now be delivered in one or at least two days from cultivation areas to the central markets of large towns. The capital city of Berlin with its many rivers, canals and lakes, used inland navigation; for example, for the sale of the highly appreciated 'Teltow root', the 'Spreewald cucumbers' and asparagus from Beelitz.[18] These kinds of fresh vegetables were sold directly out of small barges at several places in the centre of Berlin. The increasing traffic with horse-carriages on the improved main roads and provincial streets also promoted the supplying of urban consumers. But the railway remained, in speed, security and transport capacity, unbeatable until the spread of motorcycles after the First World War. The growth of traffic resulted from the enormous increase in the quantities of food supplied to urban markets. Freight costs fell and in turn lower prices extended the market. Moreover, by the reduction in time between production and sale, the capital tied up in stocks was reduced. The market-halls proved to be a great success everywhere. The numbers of exhibitors grew constantly, so that enlargements and new establishments became necessary. There the retailers and street traders supplied themselves with fruit and vegetables in the early morning and competed with the traditional women-hucksters in the weekly markets.

At first, the market-halls filled only regional and temporary gaps in the urban food provision. The mass of the urban customers stuck mostly to the weekly open-air markets which made their biggest sales on Wednesday and Saturday. The largest run was on Saturday evening when weekly wages had been paid. In addition, at some places early Sunday markets came into use, where the housewives could buy at leisure all the things necessary for the festive meal. Of course, most of the hucksters and itinerant dealers had links with peasants in the neighbourhood of the towns and could not be persuaded to move into the new market-halls. Similarly, many housewives could not get accustomed to the more anonymous market-halls. But, eventually, open-air markets lost their dominant position as urban food suppliers. It is interesting to see that the prices obtained in these markets, however, were for a long time guidelines for the sale of fresh victuals. This was due to the fact that at weekly markets so many small producers and traders met so many different customers. Therefore prices changed much faster than in the market-halls. Everybody looked to maximize profits and competition reached its optimum. But the weekly markets had to change, too. Because of the scale of some large towns a daily food market instead of a weekly one was introduced, for example, the famous *Victualien-Markt* (Market of Victuals) in Munich. Moreover, in some parts of bigger towns additional smaller weekly markets were established. In 1885 Berlin had already approximately 20, Vienna in 1912 even 40 such open-air markets.[19] By contrast, other cities like Hamburg created special markets for each main foodstuff. The opening of the famous fish market in 1911 was then followed by the wholesale vegetable market at the Deichtor, both situated next to the harbour.

The small greengrocers often set up on the ground floor of a tenement house in those town districts which were some distance from the weekly markets and market-hall. So the housewife no longer needed to go shopping to distant markets which were held twice a week and sometimes only in the early morning. It became more and more important for the urban consumer that the shopkeeper always had a large supply of potatoes, fruit and vegetables, because private households had mostly no opportunities to store large quantities of foodstuff in small and overcrowded dwellings. Furthermore, the local corner shop created a new personal confidence between the tradesman and the consumer which in addition improved the food quality. If goods were low-grade or rotten regular customers could more easily get a refund, because the shopkeeper would avoid annoying them. The possibility of shopping at the next fruit and vegetable retailer increased the power of the consumer. By putting a regular customer on an account basis the greengrocer moreover gave him a loan without interest until to the next wage-payment.[20] Not until 1900 did the first criticism arise about this "bad habit of borrowing", but appeals for cash payments increased slowly.

The 'flying street sellers' were also a new aspect of urban food provisioning. They were found at crowded street crossings and in squares as well as along access roads to markets. Initially, the towns often tried to drive these itinerant people back like the 'dishonourable' hucksters in former times, but in most cases this was impossible because the traders were highly mobile. The authorities therefore normally allowed them to sell, if they paid a small sum for a licence. Because the street-seller did not pay a rent for a shop or a fee for an open stall, and sold the cheapest kinds of fruit and vegetable from the market-halls, or even from the weekly markets, his prices were always below other groups of traders. So he was at the base of the whole competitive structure and unconsciously ensured that prices remained low.[21] Nevertheless his trading must have been profitable, as the numbers of street-sellers in Hamburg around 1900 demonstrate: In 1877 the town counted 600 itinerant traders; this number rose to 3,000 in 1912. Up to the beginning of the twentieth century the social classes with small incomes were of course the largest group of customers who bought their vegetables and fruit from the 'flying cart trader'. His daily business has been described as follows:[22]

> The fruit and vegetables which are to be sold, are bought in the earliest hours of the morning from the wholesaler or producer in the market-halls. The tradesmen drive then in the morning through their usual districts and streets. They do not mind the strain and hardship of the winter. With great regularity they stop at certain places, often even at fixed times. At the stopping point the traders ring a bell or are hawking. Then we can observe that it is not very long before the housewives are coming from the neighbourhood to the cart in order to buy vegetables and fruit for their kitchens. They save going to the market-hall, the goods are fresh and moreover cheap. Housewives who are prevented by housework, children or other circumstances from leaving the house, appreciate being able to buy their daily needs at the doorstep.

Provisioning the growing German large cities with vegetables also occurred to a limited extent by imports from foreign countries. Although we have no official national trade statistics before the foundation of the Imperial Statistical Board in 1880, we can suppose that already at the time of the German Customs Union some foreign vegetables were available at various German regional markets. During the 1848 revolution, the newly established Chamber of Commerce in Erfurt, the capital of Thuringia – a traditional centre of commercial gardening and seed production since the eighteenth century – sent a petition to the Prussian Ministry for Commerce and Trade. Their requirements were for an "exemption from duty for all vegetables and fruit in general and especially during those periods of the year when we cannot produce them. But protective duty is always there, when native products are in competition with foreign ones".[23] It was argued that the duty-free regulations should concern several kinds of cabbage, lettuce, celery, onions, cucumber, green beans, radish and mushrooms. The production of these sorts of vegetables in hot-beds and hot-houses was too expensive and therefore German gardeners could not compete with foreign countries. With the beginning of the local harvest all imports would then stop by themselves. As the Chamber of Commerce said, customers demanded, foreign vegetables only as long as insufficient quantities existed at home. Price differentials can be demonstrated from the market prices in Berlin at the same time: The price paid for home-grown green beans was 25–30 Pfennig (German pence) per pound, but for Italian ones only 10–12 Pfennig; cauliflowers from Erfurt cost 1.00–1.20 Mark, Italian cauliflowers at the same time were only 15–17 Pfennig; fresh German asparagus was 1.00–12.90 Mark, French only 40–50 Pfennig. The price survey showed that Italy was obviously already an early supplier of tomatoes, the Netherlands of cucumbers, cauliflower and carrots. On the other hand some states of the German Customs Union were regular exporters of onions and only a few imports were necessary shortly before the harvest. Before 1880 we have no exact figures for the import and export of vegetables. The already mentioned Prussian statistician Georg von Viebahn supposed in 1862 that imports of vegetables exceeded exports by far. The national statistics show, if we disregard slight deviations, from 1880 till 1913 a constant increase of vegetable imports to Germany from 25,223 tons to 308,971 tons annually.[24] In 1911 for the first time more than 100,000 tons of vegetables came from abroad. So between 1880 and 1913 imports grew more than 12 times. This underlines den Hartog's statement that the Netherlands became in this period "the greengrocer for the Ruhr area" with its huge industrial agglomerations.[25]

German vegetable exports registered a short rise between 1880 and 1884 but afterwards a rapid decrease of more than 100,000 tons. From 1885 onwards trade stagnated until the First World War at a low level. This emphasizes the views of the leading German economist Werner Sombart, who concluded from the statistics in 1909 that industrialization had caused Germany to change, as Britain had, from a traditional exporter of foodstuffs to an importing nation.[26] The segmentation of the German trade in vegetables points out that most kinds of vegetables imported came from the Netherlands, France, Italy and Austria-Hungary, with small quantities from Russia, Belgium and Spain. German vegetables on the other hand were mostly consumed to a modest degree in Austria, Switzerland and Tsarist Russia, which imported nearly all kinds of German vegetables. France was only interested in German white and red cabbage, besides onions. There was also quite a large onion export trade to all Scandinavian countries around 1900. This general balance confirms that a growing demand for vegetables very similar to that for fruit came into existence in the second half of the nineteenth century once progress into the age of industrialization and urbanization had begun.

Assessments of changes in vegetable consumption

What picture can be drawn of vegetable consumption in this period of modernization between the 1848 revolution and the outbreak of the First World War in 1914? From several statistical sources we have computed the following figures shown in Table 3.3.

Table 3.3 *Vegetable consumption per head per year in Germany 1850–1913.*

Year	kg per head	Year	kg per head	Year	kg per head
1850	37.1	1871	50.0	1892	59.4
1851	37.6	1872	50.7	1893	60.0
1852	38.2	1873	51.5	1894	60.1
1853	39.2	1874	51.7	1895	60.1
1854	39.9	1875	52.2	1896	60.3
1855	41.0	1876	52.5	1897	60.5
1856	41.6	1877	52.5	1898	60.6
1857	42.2	1878	52.8	1899	61.1
1858	42.6	1879	53.0	1900	61.5
1859	43.3	1880	51.9	1901	61.6
1860	43.9	1881	52.2	1902	61.4
1861	44.2	1882	52.8	1903	61.1
1862	44.6	1883	53.2	1904	62.3
1863	45.1	1884	53.7	1905	61.9
1864	45.7	1885	56.2	1906	61.7
1865	46.3	1886	56.5	1907	61.6
1866	46.8	1887	57.3	1908	62.4
1867	47.5	1888	57.4	1909	62.5
1868	48.0	1889	57.8	1910	62.5
1869	48.7	1890	58.4	1911	64.2
1870	49.3	1891	58.4	1912	63.9
				1913	63.5

Source: Teuteberg & Wiegelmann (1986), 236–237, 244 (graph. 8).

We can recognize here a large increase in consumption similar to the already mentioned figures of production and trade. Vegetable consumption nearly doubled from a low of 37.1 kg in 1850 to 64.2 kg in 1911. Afterwards, in the twentieth century, demand stagnated, went down during both World Wars and the time between them, with the exception of the year 1937/1938. The growth of the vegetable consumption resumed after 1960, but this time the increase was slower but steadier. By 1974/1975 consumption in the Federal Republic of Germany reached 66.7 kg per head per year which was slightly more than in 1913.

All these consumption figures suffer from the methodological problem that waste occurs at harvest, and during transport and storage. Preservation and preparing meals in the kitchen also reduces the quantities of food to a certain extent. Max Rubner, the 'Pope of the German nutritional sciences' around 1900 wrote in one of his popular handbooks: "The waste of vegetable, fruit and meat is changing and indeterminable.[27]" Later, the calculations of his colleague Wilhelm Ziegelmayer showed that all vegetable foodstuffs had varying losses of waste from 0–70 per cent or, on average, 24 per cent loss from the whole weight after the harvest. Fresh vegetables had a loss of approximately 19 per cent, when preserved 35 per cent – sauerkraut, for instance, lost 30 per cent – dried vegetables 27 per cent and pickled cucumber 20 per cent. He observed in his empirical studies that the waste of preserved winter vegetables was always much higher than the loss of home-made preserves; in the latter case waste was only 10 per cent.[28] These waste calculations would mean a reduction of vegetable consumption. On the other hand there was always a large production of green vegetables in private urban gardens which we have to add to these consumption figures. Since every second household in Imperial Germany around 1900 owned or rented a garden we guess that the home-grown produce and the losses by waste roughly compensate each other. Thus calculations of vegetable consumption per head per year can only show trends rather than exact quantities.

Conclusion

This description of the provision of vegetables for urban households between 1850 and 1913 against the background of accelerating town growth and new developments in the food trade must be valued as an attempt to throw some light on a rather obscure subject. Important aspects have only been touched on or still remained totally undiscussed. For instance, the history of vegetable preservation has been left out of consideration. The whole development of modern industrial preservation of dried vegetable soups and vegetable juices have to be left for later research. The emergence of the first factories for tinned vegetables and manufactured sauerkraut in the area of Braunschweig and Neuß in the Rhineland is well known, but never treated in a larger context. Further, an intensive historical analysis of the herbs, spices and wild edible plants, berries and mushrooms, collected in old times by women herbalists, must be of some interest now because of the running debate about the advantages of strict vegetarian nutrition and 'alternative agrarian products' or simply to overcome monotony and to look for new tastes.[29] The literature has been limited until now to botanical surveys, cooking recipes and other simple consulting writings. Likewise we do not know very much about food production from small kitchen gardens.[30] There are written sources for this and the later vegetable cultivation in urban allotment gardens. These numerous small garden-plots for the urban working classes can be traced back to the German educationalist and philantrophist Hauschild in the 1860s, who called them in remembrance of this teacher and friend Daniel Moritz Gottlieb Schreber (1808–1861) from Leipzig "Schreber gardens". The allotments, which were patronized by the municipalities for public health reason, joined together in 1921 to form a National Federation. By 1928, this league had already 2,500 local unions with 400,000 allotment gardeners.[31]

Equally, of course, we need more research about the preparation of vegetables in the kitchen and their use in meals, in relation to the progress of kitchen-technology since the last century. Books for housekeeping and cooking are a good mirror for the many distinctions between social classes and regions. As inquiries by German ethnologists convincingly demonstrate, around 1930 'food landscapes' still existed in Germany with different preferences for vegetable hot-pots and salad dishes.[32] It is interesting to find in 1970s lettuce ranked ahead of all other salads made of cabbage, cucumber, tomatoes, beans and red beets. Lettuce was normally eaten by every third family in Germany, but regional distinctions still existed.[33] A really comprehensive story of vegetable consumption would include also the questions to what extent the discovery of vitamins in vegetable plants, the search for foodstuff-substitutes during food-crises in wartime, the movement for health reforms and especially the fashionable cult for slimming in the late nineteenth and early twentieth centuries have contributed to overcome the deep-rooted scepticism against greens. Not only vegetarians, but also naturopaths preached the importance of eating more fresh fruit and vegetables, and this advice was printed constantly to a growing extent in cookery books up to the present day.

We hope this contribution makes it clear that vegetable foods have a very long history because roots and tubers belong to the oldest cultivated plants of agriculture. But the larger consumption of fresh greens developed at first in the second half of the nineteenth century due to the rationalization of traditional gardening, the liberalization of markets and the mechanization of transport. The enormous increase of vegetable consumption since then was therefore not only the result of urbanization. Though this first attempt to consider the longitudinal development is based on some rough estimates due to the waste and unknown amounts of home production, there can be no doubts that urban vegetable provisioning changed structurally and is a remarkable part of the 'revolution at the table'.

NOTES

1. Mathias (1967); Shaw (1992); Teuteberg (1987); Spiekermann (1999).
2. Teuteberg & Wiegelmann (1986), 236–241, 244, 257–261.
3. den Hartog (1996); Godina-Golija (in this volume); Teuteberg (1998).
4. Muermann (1993); Bundesausschuß Gemüse (1977); Herrmann (1969); Becker-Dillingen (1950); Kreuzer (1989); Fujino (1999),123–158.
5. See for example Bock (1539); Bock (1558); Ryff (1552); Megenberg (1861); Werneck & Speta (1980).
6. Teepe-Wurmbach (1961), 61; Tolksdorf (1971), 5–26; Bornholz (1825).
7. Bath (1963); Minchinton (1983), 76–77; Blum (1982), 81–108, see here the copper plates of early vegetable plants; den Hartog (1996).
8. Enslin (1825); Langethal (1856); Güntz (1902); Teuteberg (1989), 246–254.
9. Mehler (1798); Candolle (1824).
10. The idea of growing vegetables, fruit, flowers and exotic plants in special hot-houses during the cold months can be traced back to Roman times. Important forerunners were the numerous orangeries and winter gardens of kings and other noble families during the seventeenth and eighteenth centuries. Their gardeners either built some for commercial purposes or made detailed proposals to do so. The first hot-house using hot steam was erected in 1816 by the Swedish mechanic Martin Triwald in Newcastle-upon-Tyne using John Hoyle's patent from 1791. In Germany this

technique was not popular until the end of the nineteenth century.The first 42 hot-houses for the purpose of growing vegetables were built in the garden of the Royal Saxonian Pilnitz castle near Dresden; see Kirst-Böttner (1921).
11. Viebahn (1862), 708; see Lucas (1894).
12. Neuloh & Teuteberg (1979), 156; Teuteberg (1993), 129.
13. Fruit yields which were very profitable rose from 530,000 tons in 1850 to 2,490,000 tons in 1912.
14. Badtke (1924), 263.
15. Ibid.
16. Kaiserliches Gesundheitsamt (1905); Teuteberg, (1994).
17. Schachner (1924), 234; Hirsch (1925), 75–76; Morgenroth (1924), 332.
18. Blum (1922), 456; Lange (1911), 67–68; Samuel (1933).
19. Lehe (1966), 84.
20. Spiekermann (1999).
21. Teuteberg (1987), 17–19; Ingenkamp (1928), 121.
22. Ingenkamp (1928), 58.
23. Haupt (1908), 158.
24. Kuczinsky (1926), 89, 92.
25. den Hartog (1996).
26. Sombart (1909), 413.
27. Rubner (1913), 46.
28. Ziegelmayer (1942), 2; Ziegelmayer (1947), 671. See for the author Gassmann et al. (1999), 208–214.
29. Gramberg (1949); Winckel (1915), Lichtenstein (1972); Scheidig (1937).
30. Lüders (1775); Mehler (1798); Bergmann (1811); Nietner (1838); Bauer (1839).
31. Verein für das Wohl der arbeitenden Klassen im Kreise Waldenburg (1890); Amelung (1908); Christian (1914); *Kleingartenwacht* (1923 ff.).
32. Wiegelmann (1990), 12.
33. Institut für Absatzforschung (1970), 12–13.

BIBLIOGRAPHY

Amelung, H., *Rationeller Gemüsebau des Privatmannes im Vorort* (Berlin, 1908).

Badtke, 'Lebensmittelversorgung', in: Josef Brix et al. (eds), *Handwörterbuch der Kommunalwissenschaften* 3 (Jena, 1924), 262–264.

Bath, Slicher van, *The Agrarian History of Western Europe, 500–1859* (London, 1963).

Bauer, Jacob, *Der Küchengartenfreund; oder, Anleitung zur vortheilhaften Benützung der Gemüsegärten* (Carlsruhe, 1838).

Becker-Dillingen, Josef, *Handbuch des gesamten Gemüsebaus. Einschließlich der Gewürz- und Küchenkräuter* (1924), 5th edn (Berlin / Hamburg, 1950).

Bergmann, J. G., *Deutschlands wildwachsende Nahrungsmittel, I. Abtl.: Surrogate für Getreidebrote, Gemüse, Salate, Gewürze* (Breslau, 1811).

Blum, 'Güterverkehr', in: Josef Brix et al. (eds), *Handwörterbuch der Kommunalwissenschaften* 2 (Jena, 1922), 456–457.

Blum, Jerome, *Our Forgotten Past: Seven Centuries of Life on the Land* (London, 1982). German: *Die bäuerliche Welt. Geschichte und Kultur in sieben Jahrhunderten* (München, 1982).

Bock, Hieronymus, *Kleines Kräuter-Buch* (Straßburg, 1539).

Bock, Hieronymus, *Teutsche Speißkammer…* (Straßburg, 1558).

Bornholz, A., *Der Trüffelbau* (Quedlinburg, 1825).

Bundesausschuß für volkswirtschaftliche Aufklärung (ed.), *Gemüse* (s. l. s. a. / Bonn, 1977).

Candolle, Adolphe Pyramide de, *Die verschiedenen Arten, Unterarten und Spielarten des Kohls und der Rettige, welche in Europa erbauet werden. Aus dem Französischen von C. F. W. Berg* (Leipzig, 1824).

Christian, *Städtische Freiflächen und Familiengärten* (Berlin, 1914).

Enslin, Theodor Christoph Friedrich, *Bibliotheca oeconomia oder Verzeichniß aller brauchbaren, in älterer und neuerer Zeit bis zur Mitte des Jahres 1824 in Deutschland erschienenen Bücher über Land- und Forstwirthschaft…* (Landsberg a. d. W., 1825).

Fujino, Y.: 'Food characteristics of vegetables'. *Ernährungsforschung* 13 (1999), 123–158.

Gassmann, Berthold et al., 'Zur Geschichte der institutionalisierten Ernährungswissenschaft in Deutschland'. *Ernährungs-Umschau* 43 (6) (1996), 208–214.

Gramberg, E., *Wildgemüse, Wildfrüchte, Haustee*, 5th edn (Heidelberg, 1954).

Güntz, Max, *Handbuch der landwirtschaftlichen Literatur*, 2 vols (Leipzig, 1897–1902).

Hartog, Adel P. den, 'Vegetables as Part of Dutch Culture: Invention or Tradition?', in: Hans-Jürgen Teuteberg & Gerhard Neumann & Alois Wierlacher (eds), *Essen und kulturelle Identiät – Europäische Perspektiven, Kulturthema Essen 2* (Berlin, 1996), 372–386.

Haupt, Hans, *Die Erfurter Kunst- und Handelsgärtnerei in ihrer geschichtlichen Entwicklung und wirtschaftlichen Bedeutung dargestellt* (Jena, 1908).

Herrmann, Karl, *Gemüse und Gemüsedauerwaren* (Berlin / Hamburg, 1969).

Hirsch, Julius, *Der moderne Handel, seine Organisationen und Formen und die staatliche Binnenhandelspolitik*, 2nd rev. edn (Tübingen, 1925).

Hoffmann, Walter G.[ustav], *Das Wachstum der deutschen Wirtschaft seit der Mitte des 19. Jahrhunderts* (Berlin / Heidelberg / New York, 1965).

Ingenkamp, Franz, *Der großstädtische Straßenhandel*, Diss. rer. pol. (Köln, 1928).

Institut für Absatzforschung (ed.), 'Da hat man den Salat'. *Ifak-Dienst* 28 (1970), 12–13.

Kaiserliches Gesundheitsamt (ed.), *Gesundheitsbüchlein. Zusammengefaßte Anleitung zur Gesundheitspflege*, 11th edn (Berlin, 1905).

Kirst-Böttner, *Gewächshäuser*, 2nd edn (Berlin, 1921).

Kleingartenwacht (Berlin, 1923 ff).

Kreuzer, Johannes, *Gartenpflanzen-Lexikon 3: Obst, Gemüse, Kräuter*, 2nd edn (Braunschweig, 1989).

Kuczinski, René Robert, *Deutschlands Versorgung mit pflanzlichen Nahrungs- und Futtermitteln* (Berlin, 1926).

Lange, Edgar, *Die Versorgung der großstädtischen Bevölkerung mit frischen Nahrungsmitteln unter besonderer Berücksichtigung des Marktwesens der Stadt Berlin* (Berlin, 1911).

Langethal, Christian Ernst, *Geschichte der teutschen Landwirtschaft*, 4 vols (Jena, 1856).

Lehe, Erich von, *Die Märkte Hamburgs von den Anfängen bis zur Neuzeit*, 1st edn 1911 (Wiesbaden, 1966).

Lichtenstein, H., *Wildgemüse, Wildsalate, Hausteepflanzen, Gewürzpflanzen* (München, 1972).

Lucas, E., *Der Gemüsebau*, 5th edn (Stuttgart, 1894).

Lüders, F. H. H., *Die in Briefen über die Bestellung des Küchengartens vorkommenden Küchengartengewächse* (Leipzig, 1775).

Mathias, Peter, *Retailing Revolution* (Oxford, 1967).

Megenberg, Konrad, *Das Buch von der Natur*, ed. by Fr. Pfeifer (Stuttgart, 1861).

Mehler, Johann, *Die Landwirthschaft des Königreichs Böhmen* 4, 1: *Von den Küchengärten* (Prag / Dresden, 1798).

Minchinton, Walter, 'Die Veränderungen der Nachfragestruktur von 1500 bis 1750', in: Carlo M. Cipolla & Knut Borchardt (eds): *Europäische Wirtschaftsgeschichte, vol. 2: Sechzehntes und siebzehntes Jahrhundert* (Stuttgart/ New York, 1983).

Morgenroth, Wilhelm, 'Markthallen II', in: *Handwörterbuch der Kommunalwissenschaften* 3 (Jena, 1924), 339–345.

Muermann, Bettina, *Lexikon Ernährung*, 2nd edn (Hamburg, 1993).

Neuloh, Otto & Teuteberg, Hans-Jürgen, *Ernährungsfehlverhalten im Wohlstand. Ergebnisse einer empirisch-soziologischen Untersuchung heutiger Familienhaushalte*

(Paderborn, 1979).
Nietner, Theodor, *Die Küchengärtnerei*, 2 vols (Berlin, 1838).
Rubner, Max, *Wandlungen der Volksernährung* (Leipzig, 1913).
Ryff, Walter, *Kleines Teutsches Apoteck oder Confect- und Latwergen-Büchlin...* (Straßburg, 1552).
Samuel, L., *Gemüse, Obst und Südfrüchte im Deutschen Reich. Versorgungsbilanz und Verkehrsbeziehungen* (Berlin, 1933).
Schachner, Richard, 'Markthallen I. Technisches', in: Josef Brix et al (eds), *Handwörterbuch für Kommunalwissenschaften* 3 (Jena, 1924), 333–339.
Scheidig, K., *50 deutsche Wildpflanzen als Salate, Gemüse und Rohkost* (Leipzig, 1937).
Shaw, Gareth, 'Large-scale retailing in Germany and the development of new retail organizations', in: John Benson & Gareth Shaw (eds), *The Evolution of Retail Systems, c. 1800–1914* (Leicester / London / New York, 1992), 166–185.
Sombart, Werner, *Die deutsche Volkswirtschaft im neunzehnten Jahrhundert*, 2nd rev. edn (Berlin, 1909), 7th edn (Berlin, 1927).
Spiekermann, Uwe, *Basis der Konsumgesellschaf.: Entstehung und Entwicklung des modernen Kleinhandels in Deutschland 1850–1914*, (München, 1999).
Teepe-Wurmbach, A., 'Kohl – Mus – Kraut. Wort– und sachkundliche Untersuchungen zur norddeutschen Gemüse- und Obstbereitung'. *Westfälische Forschungen* 14 (1961), 150–168.
Teuteberg, Hans-Jürgen & Wiegelmann, Günter, *Unsere tägliche Kost. Geschichte und regionale Prägung*, 2nd edn (Münster, 1986).
Teuteberg, Hans-Jürgen (ed.), *Durchbruch zum modernen Massenkonsum. Lebensmittelmärkte und Lebensmittelqualität im Städtewachstum des Industriezeitalters* (Münster, 1987).
Teuteberg, Hans-Jürgen, 'Reise- und Hausväterliteratur der frühen Neuzeit', in: Hans Pohl (ed.), *Die Bedeutung der Kommunikation für Wirtschaft und Gesellschaft* (Berlin, 1989), 216–254.
Teuteberg; Hans-Jürgen, 'Prolegomena zu einer Kulturpsychologie des Geschmacks' in: Alois Wierlacher & Gerhard Neumann & Hans Jürgen Teuteberg (eds), *Kulturthema Essen. Ansichten und Problemfelder* (Berlin, 1993), 103–136.
Teuteberg, Hans J., 'Food adulterations and the beginnings of uniform food legislation in late nineteenth-century Germany', in: John Burnett & Derek J. Oddy (eds), *The Origins and Development of Food Policies in Europe* (Leicester / London / New York, 1994), 146–160.
Teuteberg, Hans-Jürgen, 'Obst im historischen Rückspiegel – Anbau, Handel, Verzehr'. *Zeitschrift für Agrargeschichte und Agrarsoziologie* (1998), 168–199.
Tolksdorf, Ulrich, 'Pilze als Nahrung. Zu Vorurteil und Innovation eines Nahrungsmittels in Norddeutschland'. *Kieler Blätter zur Volkskunde* 3 (1971), 5–26.
Verein für das Wohl der arbeitenden Klassen im Kreise Waldenburg (eds), *Ueber die Vortheile des Gärtnereibetriebes, nebst einer Anweisung zum Anbau und zur Verwerthung von Beeren, Gemüsen, technischen und offizinellen Pflanzen sowie einer kurzen Anleitung zur Obstbaumkultur* (Waldenburg i. Schl., 1890).
Viebahn, Georg von, *Statistik des zollvereinten und nördlichen Deutschland* 2 (Berlin, 1862).
Werneck, Heinrich L. & Speta, Franz (eds.), *Das Kräuterbuch des Johann Hartlieb. Eine Bilderhandschrift aus dem 15. Jahrhundert* (Graz, 1980).
Wiegelmann, Günter, 'Historische Grundlagen der regionalen Kostunterschiede in Mitteleuropa', in: Siegrid Weggemann (ed.), *Alte Landschaftsküchen in neuer wissenschaftlicher Bewertung* (Frankfurt am Main, 1990), 4–15.
Winckel, R., *Wilde Gemüse* (Magdeburg, 1915).
Ziegelmayer, Wilhelm, *Unsere Lebensmittel und ihre Veränderungen. Mit einer Darstellung der Lehre von der Kochwissenschaft*, 3rd rev. edn (Dresden / Leipzig, 1942).
Ziegelmayer, Wilhelm, *Die Ernährung des deutschen Volkes*, 5th rev. edn (Dresden / Leipzig, 1947).

ADEL P. DEN HARTOG

4 Vegetables for Dutch townspeople

The role of marketing and nutrition education, 1870–1990

Vegetables are now an essential part in the diet of western Europe though looking at present food consumption patterns, north-west European countries consume fewer vegetables compared with the south, in particular, the Mediterranean region.[1] Vegetables are considered to be a necessary part of the Dutch diet, based on a long tradition of relatively high consumption. In the nineteenth century, however, the consumption of vegetables was low, in particular among the lower-income classes. A similar situation can be found in other European countries as well. It is often said that the rise in the standard of living has led to an increase in the consumption of relative expensive food products. However, not much is known of the social and economic processes behind the steady increase in consumption of foods such as vegetables. The aim of this chapter is to analyze how and why vegetables have become an essential part in the diet. Three factors have played a decisive role in the process of diffusion and acceptance of vegetables: the marketing activities of the horticulture sector, the teaching of schools of home economics and the influence of nutrition education.

Vegetables in the European context

At present the level of vegetable consumption varies significantly between the countries of the European Union (see Table 4.1). The vegetable consumption of northern Europe is considerably lower than southern Europe. France lies in between. This is not surprising as, according to geographers, France has its feet in the Mediterranean and its head in the north. Until the 1980s the Netherlands, with its important horticulture sector, had the highest vegetable consumption among the countries of north-west Europe. Now Belgium has taken over this position.

Table 4.1 *Availability of vegetables for human consumption in the European Union in kg per caput per year.*

Belgium/ Luxembourg	110.9 (1996/1997)	Sweden	63.1 (1996/1997)
UK	96.4 (1995/1996)	Finland	61.4 (1996/1997)
Netherlands[1)]	93.8 (1994/1995)	Greece	308.3 (1994/1995)
Ireland	90.4 (1996/1997)	Italy	174.5 (1995/1996)
Austria	90.3 (1996/1997)	Spain	168.2 (1996/1997)
Germany	88.1 (1996/1997)	Portugal	125.3 (1996/1997)
Denmark[2)]	n.a.	France[3)]	n.a.

Source: Eurostat.

Note: [1)] Netherlands in 1993/1994 118.5 kg.
[2)] Denmark in 1987 60.0 kg.
[3)] France in 1987 113.0 kg.

In the Mediterranean region the consumption of fresh and in particular uncooked vegetables has a long tradition. In northern Europe the position of vegetables in the diet has been less clear.[2] Is there an ecological factor involved? The growing season in north-west Europe is shorter than in southern Europe. When people do not consume certain foods there is, according to Marvin Harris, always logic behind their choices. In countries or regions which lie in the north the ecological conditions for horticulture are of course less favourable than in southern countries. However, cultural factors also play a role. Thouvenot has demonstrated clearly that within the same geographical area cultural differences exist.[3] In the north-east of France people who live to the west of the Romance-Germanic language frontier eat their *chou rouge* or *Rotkohl* raw in the form of a salad, and east of the language frontier only in cooked form.

In the Netherlands, as well as in other countries, the appreciation of vegetables was limited in the Middle Ages.[4] Eating raw food was considered unhealthy.[5] It is important to note that the old Dutch word for vegetables is *warmoes*, a contraction of the words *warm* and *moes* which is related to the old low German word *warmôs*; or to say it in English, warm and vegetable. It confirms the concept that vegetables should be cooked. For the common people vegetables were not used as an accompaniment to the main course of the meal, which among the gentry consisted of meat or fish. A stew or hotchpotch (*hutspot, potage* or *stoofpot*) based on coarse vegetables, cabbage, carrots, turnips, onions and leek was a main dish of the common people.[6] Cereal porridges, pancakes and stews remained the main dishes of Western Europe well beyond the Middle Ages.[7] The diversification of the diet was stimulated by the establishment of the Dutch Republic and the rise of a town-based merchant class in the sixteenth and seventeenth centuries. It was then that the more well-to-do in the towns started to make more use of vegetables.[8]

Vegetable production and food habits

In the nineteenth century the appreciation of vegetables in the Netherlands was limited. The eating of uncooked or raw vegetables was still largely rejected. Budget surveys in the period 1850–1914 indicate that the diets of both urban and rural working-class households were mainly based on potatoes and, at the end of the nineteenth century, on potatoes and bread. The most commonly used vegetables were turnips, carrots and onions. They were cooked with potatoes and a little fat as a casserole or one-dish meal. In the summer French or green beans were used and cooked apples and pears as autumn vegetables.[9]

Between 1873 and 1880 the Royal Netherlands Medical Society initiated a number of studies on the nutritional status of the working class. Ten reports were published which showed that the labourers' menu was based on potatoes; vegetables were hardly included in the diet.[10] The finer vegetables were considered as a luxury and those who could have spent some money did not do so, because they thought vegetables would not fill an empty stomach.[11] There was a basis of truth in this statement. More vegetables can be consumed when the energy density of food increases, as the diet becomes less bulky. The increase of fat in the total daily energy intake, estimated at 17 per cent for labourers in 1907 and 38 per cent in the 1990s made it physiologically possible to consume more vegetables, which were foods with a high water content.[12]

The demand for more and finer vegetables emerged in the second part of the nineteenth century, as a steady growth of cities, with higher income classes because of the industrial revolution, occurred especially after 1870.[13] Among the upper classes a dietary evolution took place. The *hutspot* or casserole evolved into a dish based on three separately cooked components, a 'triad' consisting of potatoes, meat and vegetables.[14] Thus vegetables arrived on the plate. The development of what is now considered to be the traditional Dutch main dish, the 'triad' of potatoes, meat and vegetables was associated with profound changes in the cooking techniques. In the nineteenth century the lower classes had only one cooking place at their disposal, the open fireplace or hearth. In the second part of the nineteenth century industrially produced cast-iron cookers became increasingly affordable. This meant that several dishes could be cooked at the same time. Enclosing the fire was a relatively rapid process. Around 1850 most households still had an open fireplace; at the beginning of the twentieth century the open fireplace was already an exception. The last open fireplaces in the Netherlands disappeared around 1940 in the east of the country.[15]

In the Netherlands professional horticulture emerged in the province of Holland in the area surrounding the cities Delft and Leiden.[16] For example, in 1581 the city of Leiden, with a population of about 12,000 inhabitants, contained 51 self-employed vegetable growers or market gardeners. The gardens were just outside the city ramparts.[17] In the following century the more well-to-do in the towns were starting to consume more vegetables and this laid the foundation of horticulture as a major economic activity in the

Netherlands, not only in the field of fruit and vegetables but likewise in ornamental plants. The increasing frequency of vegetable consumption in the course of the centuries are presented in Table 4.2. In the eighteenth century a change took place in the production of vegetables and their consumption among the more well-to-do in the towns, a shift from coarse vegetables to finer vegetables. The coarse and relative cheap vegetables remained in the diet of the common people: cabbage, turnips, carrots, onions and beets. The newer more delicate vegetables for the higher social classes were cauliflower, endive, cucumber, celery and chervil.[18]

Table 4.2 *Estimated number of main kinds of vegetable consumed in the Netherlands between the thirteenth and twentieth centuries[1)].*

Period	Number	Period	Number
13th century	5	18th century	18
14th century	7	19th century	24
15th century	9	1956	25
16th century	12	1990	57
17th century	15	1996	63

Source: Based on Sangers (1963); Nederlandse Voedingsmiddelentabel (1990, 1996).
Note:[1)] Herbs and aromatic plants are not included.

In the middle of the nineteenth century the production and variety of vegetables started to increase, particularly after 1870. The demand for vegetables was stimulated by two major developments, the earlier mentioned rapid urbanization of the Netherlands and the export trade to London and the relatively nearby Ruhr area in Germany.[19] In a country well endowed with rivers and canals, it was not surprising that cities were supplied with vegetables by means of barges and to a lesser extend by trains. Thus, the connection of the Dutch railway system with that of Germany in the 1870s was a major factor in supplying the cities of the Ruhr area with Dutch vegetables. Cultivation under glass was introduced on a large scale, first in garden-frames and later in hot-houses. This extended the growing season considerably.[20] It marked the beginning of a process to make fresh vegetables available throughout the year and a near disappearance of eating according to the season. At the beginning of the twentieth century the horticulture of the Netherlands was technically highly developed and strongly export oriented.

Reaching the urban consumer

In pre-industrial Europe farmers sold their produce in town markets. Several street names in Dutch cities bear witness to farmers' activities (The

Groenstraat, and Warmoesstraat in Amsterdam, Coolsingel in Rotterdam). At the end of the nineteenth century two main systems prevailed. First, the sale of vegetables from areas near the towns was mainly done by the market gardener himself. Sellers and buyers met at the city market where the buyers could be hawkers, greengrocers and consumers. Second, the vegetables produced in areas far away from the cities were bought by middlemen.[21]

As demand for fresh vegetables and fruit increased, this proved to be a fragmented and inefficient system. In such a system the market gardener could not get the best possible price for his product. A market gardener gave a description of the situation of the horticultural area of the Westland round 1870, which can be summarized as follows:

> With good or bad weather, deep in the night the small towing barge with vegetables went to the vegetable market of The Hague. At the market though negotiations took place with wholesale buyers with their unreasonable demands. Direct sale to customers was done at the doorstep by the vegetable gardeners' boys. With due respect and cap off, they showed the content of the vegetable basket to the maid, hoping she had the goodness to buy something. The output of a hard working family of market gardeners depended on the whims of middlemen and maids.[22]

In 1887 the first auction market was organized where vegetable farmers could sell their produce by auction to middlemen, hawkers and greengrocers.[23] Other auctions followed; the vegetable auctions remained in the hands of the producers' associations formed by horticulturists. In 1913, 22 vegetable auctions decided to work closely together in the interest of the whole horticulture sector: vegetables, fruits and ornamental plants.[24] The Central Bureau of Vegetable Auctions in the Netherlands (*Centraal Bureau van de Tuinbouw Veilingen in Nederland*) became very influential in the process of diffusion and acceptance of vegetables among the Dutch population, and the promotion of exports.

The Central Bureau was in fact a co-ordinating agency of vegetable auction associations with its office in The Hague. In 1924 the board proposed to its members the creation of an advertising fund. The board stated that advertising was essential since a product could no longer be sold only on the merit of its quality. Not all members were enthusiastically in favour of this idea and the board had to solicit for the necessary support.[25] The idea was accepted and, on a voluntary basis, each member association contributed 0.1 per cent of its yearly turnover to the fund. Most members did.[26] The fund was used for advertising activities both in the Netherlands as well as abroad, in particular in Germany and the United Kingdom. The creation of the Advertising Fund in 1924 marked the start of a long lasting collective advertising and promotion for vegetables and fruits for the rest of the twentieth century. It became the first important factor in the process of vegetable diffusion as a healthy food. The Central Bureau adopted the concept of collective advertising from the United States and the United Kingdom. They were acquainted with the intensive collective advertising done by vegetable producers and traders in these countries. Until then, advertising in the Netherlands was only used for proprietary brands.[27]

Advertising of vegetables and nutritional sciences, 1924–1940

In 1924 the advertising campaign started with a free distribution of 250,000 tomato recipe booklets and 7,500 posters to shopkeepers.[28] The tomato was chosen as a kind of a test case for the Advertising Fund. Tomato consumption was considered to be too limited. The first commercial tomato cultivation in the Netherlands took place in 1897, in two new greenhouses in the centre of the horticulture district of Westland (Poeldijk). Before that time tomatoes were cultivated in gardens of estates or in domestic gardens, mainly as ornamental plants because of the beautifully coloured red and orange fruits. They were rarely sold.

In the beginning, sales of commercially produced tomatoes were disappointing. Some tomato baskets went through the auction system to luxury shops but did not reach a large public. In this period exports to Germany and the United Kingdom were more important than home consumption. Gradually tomatoes began to be accepted by Dutch consumer, so the tomato was a target for a collective advertising campaign. The Central Bureau stated that the campaign was in the interest of producers as well as the vegetable and fruit traders. They thought it would not be too difficult to make the Dutch public better acquainted with the tomato.[29] The Central Bureau introduced a clear nutritional basis for the campaign. They concluded that the way of life in the Netherlands had changed, with hundreds of thousands of people leading sedentary lives in offices, where no muscular force or physical activities were needed. For them health required a different nutritional message, a turning away from meat and lard and a greater use of vegetables and fruit. The aim of the advertising campaign was to change food behaviour for better public nutrition, and for the benefit of the producers. On the first poster issued by the Bureau was clearly presented the slogan: "Eat more tomatoes, healthy for young and old".[30] The campaign was considered to be successful, as home consumption of tomatoes increased.[31] Based on these results the Bureau increased its advertising activities. In the period 1927–1940 the diffusion of vegetables and fruit as healthy products was done by means of the following approaches:

Free distribution of recipe booklets

The free distribution of recipe booklets to consumers by means of shops and exhibitions was a major element of the advertising activities. The tomato booklet was too specific and to reach a larger public a more general recipe booklet on fruits and vegetables was issued in 1926, with a circulation of 400,000 copies. In 1937 it was revised by a teacher of the school of home economics in Rotterdam, Miss E. Plate.[32] The fruit and vegetable recipe booklet became much appreciated by housewives and was used in schools of home economics.[33]

Posters

Based on the successful acceptance of the tomato poster by shopkeepers as well as consumers, a more general poster was introduced in 1928. The poster had a well presented slogan "eat more vegetables and fruit, they provide health, strength and joy of living".[34] In 1929 the Central Bureau came to the conclusion that posters were indispensible for advertising. Besides the necessity to remain in touch with the customers, there was also a clear danger coming from advertising made for Mediterranean and other imported fruits.[35] Posters were also placed in station halls and on railway platforms. In 1928 transparent posters of vegetables and fruits were put in trams in the major cities.[36] A related activity had been the supplying of paper bags to shopkeepers in 1929. On the paper bag was printed the earlier mentioned slogan, "eat more vegetables and fruits; they provide health, strength and joy of living".[37]

Other means

The Bureau understood the importance of new media, such as radio and film. In 1926 the various professional horticultural organizations managed to get a weekly half-hour programme on Hilversum radio.[38] In 1928 the Central Bureau showed a horticultural film at an exhibition for the first time. It was a rather expensive medium, but the Central Bureau considered it as a more lasting means of advertising.[39] In 1931 a tomato film underlining the importance of the tomato as a healthy food was shown in major cinemas.[40] From 1929 onwards efforts were made to get illuminated signs with slogans to eat more vegetables and fruits in quality greengrocers' shops. The idea was that the illuminated sign would attract consumers even when the shops were closed.[41]

Advertising on billboards was discontinued in 1932. The effect of posters on fruit and vegetables amidst posters on other products was seen as ineffective. A new means was chosen, the advertising car. Four cars decorated with advertisements for vegetables and fruits roamed the major cities of the country handing out recipe booklets at busy points.[42] However, one important medium for food advertising, women's magazines and newspapers, was hardly used by the Central Bureau.

During the whole period 1927–1940, the Central Bureau maintained the nutritional basis of its advertising activities. This grew in importance during the economic crisis of the 1930s. The economic depression of 1929 hit the export-oriented Dutch agriculture heavily. Exports were frustrated by increasing import restrictions and in the home market demand fell due to unemployment and decreasing purchasing power. Horticulture was compelled to place more emphasis on the home market. In 1931 the Central Bureau observed that vegetables and fruits were consumed by all classes of the society but there was still room to increase consumption (see Table 4.3). Fortunately new work by scientists pointed out the importance of vegetables and fruits for human consumption. Scientific experiments had proved that salts and vitamins indispensable for human growth, were all to be found in vegetables and fruits. Needless to say the Central Bureau found such propaganda from an unsuspected source most welcome.[43] This reinforced the nutritional dimensions of their advertisements. The new tomato campaign of 1931 had as slogan on its poster: "tomato time means health". Despite financial constraints on the horticultural sector in a period of economic crisis, contributions to the Advertising Fund were increased. Advertising had to take into account not only the importance of vitamins, but also a new trend: the consumption of raw vegetables as salads. For all products the threefold "strength, health, joy of life" had to be presented.[44]

Table 4.3 *Consumption of vegetables and other foods in labour households per consumer unit a day in g in 1916 and 1929/1930.*

Food	1916[1)]	1929/1930[2)]
Potatoes	376	312
Meat/fish	41	83
Vegetables	159	153
Bread	446	369
Milk	303	655
Fat	40	75

Source: Sajet & Polak (1916); Banning (1931).

Note: [1)] Labour households in Amsterdam.
[2)] Labour households in Zaandam.

From 1935 onward horticulture was plagued with overproduction to such an extent that products had to be withdrawn from the market. Apart from destroying the unsold products the Central Bureau looked for better alternatives. In 1936 a start was made with the distribution of unsold vegetables and fruits to the unemployed and destitute households, under guidance of the Minister of Social Affairs. The government also supported the preservation of unsold products in tins during the high season for the purpose of distributing these tinned vegetables to poor households during winter time.[45] This food distribution was generally not well received by the beneficiaries, mainly because of the indignity of the system.[46] The problem of unsold tomatoes in 1938 and 1939 was solved by government support. The tomatoes were processed into tinned tomato soup and sold to the public at a low price accompanied by an intensive advertising campaign. All the tins were sold.[47]

The economic crisis made the professional horticultural sector suspicious of new developments by townspeople and country folk in their allotment gardens. In 1936 the Central Bureau observed an increase of complaints about competition from cheap vegetables and fruits grown on allotments. The increase in acreage and sale of products was alarming. Another concern was a possible trend towards self-sufficiency amongst town-dwellers, which meant a loss of consumers. On the other hand the allotment as a pastime could not be challenged. The Central Bureau sent an official letter to the Minister of Agriculture and Fisheries requesting appropriate measures.[48] The

Minister responded that he was willing to monitor the vegetable production of allotments and that he had the legal means to take appropriate action in case of any illegal trade.[49] The conflict showed that during a severe economic crisis town-dwellers would attempt to achieve self sufficiency in food, particularly vegetables and fruit. This process was reinforced during the occupation of 1940–1945.

Vegetable promotion, continuity and change 1945–1990

In the period of reconstruction after 1945, leaders of the horticultural sector became convinced that promotion of vegetable consumption, both at home and abroad, was absolutely necessary for the future of their industry. Since the 1950s two organizations have become actively involved in vegetable promotion activities: The Central Bureau of Vegetable Auctions, and The Interprofessional Organization for Vegetables and Fruit (*Produktschap voor Groenten en Fruit*), created in 1954.

In the years 1950–1970 both organizations were involved in large scale campaigns directed at consumers and in particular housewives. The promotional activities followed largely the approaches already established before 1940, though some ideological changes did occur. In 1949 the terms advertising and propaganda were still used.[50] From 1951 onwards the term propaganda disappeared and advertising was gradually replaced by education of the public (*voorlichting*).[51] Probably the term education was introduced because of close contacts with the Bureau of Nutrition Education. After all, vegetables and fruits were highly recommended by the public health sector. Education sounded more objective in contrast to advertising and propaganda. The vegetable and food promotional activities included:

- Participation in consumer exhibitions;
- Extension meetings with women's organizations;
- Education materials for schools;
- Films, slides, folders and brochures.

The core of the message was that vegetables form a tasty, healthy and indispensable part of the diet. Close contacts were maintained with the Bureau of Nutrition Education. In this period no collective advertising campaigns were initiated for vegetables, except for sauerkraut.[52] The above mentioned activities were however nation-wide. A very popular promotional film in the 1960s was called *Mijn liefde is groen* (My love is green).[53] In 1951 the Central Bureau initiated an informative radio programme on vegetables directed to housewives. After 12 years this programme was still popular. It provided information on availability and prices of vegetables on a daily basis, together with suggestions for preparation.[54] This was a remarkable aspect of free publicity, as commercials on radio and television were not allowed in the Netherlands until 1967. The spread of promotional material through retail outlets and exhibitions was extensive. In 1959 for instance 1.5 million folders were distributed on a monthly basis.[55] A nation-wide campaign started in

1970 using the slogan "fresh is healthy" with the aim of promoting the consumption of fresh vegetables.[56] The recipe booklet *vers is gezond* (Fresh is healthy) soon became a bestseller: 200,000 copies were sold. The introduction of relatively new vegetables into the ordinary diet was promoted. According to the Central Bureau, vegetables such as paprika or sweet pepper, mushrooms and asparagus were still unknown by many housewives. So, in 1966, a promotional campaign was initiated.[57] In the 1980s special promotional material was prepared for vegetables such as iceberg salad, Chinese cabbage, courgettes or marrow, fennel and broccoli. Both organizations further promoted the habit of eating raw vegetables and salads by means of folders and recipe booklets.[58]

Training in home economics for girls

Various authors have argued that cookery classes and schools of home economics for girls contributed to a spread of modern nutritional thinking among the population.[59] At the end of the nineteenth century, schools of home economics were established for girls, mainly of the higher social classes. Later, schools of home economics followed for rural girls coming from small scale farms and agricultural labourers' households. These schools were funded from private sources. From 1921 onwards schools of home economics were provided by the state following the new Vocational Training Act. This meant that training in home economics became much more accessible for girls from the lower classes.[60] Hygiene and nutrition were major components of the curriculum. In the 1920s, after the discovery of vitamins, more emphasis was put on vegetables, particularly fresh vegetables. The use of uncooked vegetables was encouraged, too.[61]

As far as vegetables were concerned the practical and theoretical lessons comprised the following elements:

- The hot main meal of the day should consist of the three components: potatoes, vegetables, and some meat (or fish). Teachers played a crucial role in getting this triad accepted among the population.
- Variation in the diet, which always included vegetables and fruit.
- Concern to avoid the loss of vitamins. Only a small amount of water was used for cooking, and cooking times were shortened. Simmering of potatoes and vegetables was considered old-fashioned. "Mind the vitamins" became a well-known expression.
- Introduction of the concept of raw vegetables as part of a meal. In the rural schools of home economics, horticulture and home preservation of vegetables were part of the curriculum.[62] In the 1960s the schools of home economics started to decline because of better education opportunities for girls and women's emancipation. In the period 1920–1950, however, schools of home economics had contributed to a further spread of the concept of vegetables as part of the diet.

Nutrition education

The gradual acceptance of vegetables and fruits as a major contribution to health started with the discovery of vitamins at the end of the nineteenth and the beginning of the twentieth centuries. Research in the 1920s revealed that nearly all vegetables are rich in carotene (pro vitamin A) and vitamin C, and contain significant amounts of calcium, iron, and other minerals. The amount of the B vitamins, however, is small. Nutrition research was dominated by vitamin issues until the 1940s. However, long before the establishment of nutrition education, the horticultural sector had been emphasizing the importance of vegetables and fruit for health to the general public.

In the Netherlands nutrition education directed at the general public started during the depression as part of the growth of home economics extension classes in the middle of the 1930s. Two organizations were set up for home economics extension teaching, one for the urban and one for the rural areas. Home economics extension education was focussed on low income households, and nutrition was a major component. Their nutritional message can be summarized as follows:

- A diet should always include vegetables.
- Food preparation should be economical, with a minimum of food waste.
- Cooking time should be short to save fuel and to prevent loss of vitamins.[63]

It is obvious that the nutrition message was largely the same as the nutrition part of the curriculum of the schools of home economics. The extension class teachers came from the schools of home economics. Reducing the cooking time of vegetables and potatoes in order to reduce the loss of vitamins was a recurrent theme in extension activities.

Nutrition education was further enhanced with the creation of a national Bureau of Nutrition Education in 1941. The Bureau became the centre of nutrition education and overshadowed the home economics extension movement. In 1953 the Bureau of Nutrition Education introduced the so-called food guide as an instrument for nutrition education. It depicted a circle with five segments. Each segment presented a group of foodstuffs with a specific and indispensable contribution to the recommended nutrient intake. Potatoes, vegetables, and fruit formed one segment in the circle. The idea behind this concept was to make housewives think in terms of foodstuffs and not in nutrients. The food guide was based on American experiences with the "basic seven [segments] of the wheel of good eating". Vegetables retained their importance in nutrition education and even improved it. In 1981 the five segments were reduced to four, when potatoes were moved to the segment of staple cereal foods. Thus the group of vegetables and fruits became even more clearly visible.[64]

The importance of vegetables in the diet was further emphasized by the dietary goals set by the Nutrition Council of the Netherlands in 1986. The core of the dietary goals was a reduction of the consumption of animal foods and fat in general[65] and an increase in the consumption of foods of vegetable origin. The dietary goals recommended an increase in the consumption of more carbohydrate and dietary fibre. The practical advice given by the Bureau of Nutrition Education was: eat more vegetables and use them also as a spread in sandwiches. Needless to say, Dutch horticulture was very pleased with this way of nutritional thinking. It is interesting to note that nutritionists have always appreciated vegetables more than meat for nutritional well being. For a long time vegetables were appreciated because of their vitamin contributions to the diet and in the 1980s for their dietary fibre. In Europe as well as in the Netherlands cardiovascular disease and cancer became major causes of disease death (40 per cent and 28 per cent of deaths, respectively). In the 1990s epidemiological studies indicated that high consumption of vegetables and fruits was likely to reduce risks of cancer, and to a lesser extent the risk of cardiovascular disease.[66] This was welcome news for the horticultural sector faced with a trend of a diminishing vegetables and fruit consumption in the 1990s. They set themselves the target of achieving an increase in daily vegetable consumption to 200 g and 2 pieces of fruit. However, in the years 1988–1992 the mean daily vegetable consumption decreased from 144 to 128 g. In 1995 a nation-wide campaign was started to promote vegetables and fruit for better health,[67] though this may be threatened by environmentalists challenging the 'unhealthy' ways of producing vegetables. Related to this are discussions on the Acceptable Daily Intake (ADI) of nitrate because vegetables have a relatively high nitrate content, particularly those cultivated in the winter period.

Conclusion

The horticultural sector was able to supply fresh vegetables and fruits to a rising urban population. Thanks to the innovativeness of the sector and imports from other geographical zones, urban consumers largely lost the habit of eating according to season. A coincidence of economic and nutritional (public health) interests can be an important factor in changing food habits. This does not mean, however, that this change will fully sink in. Those in the Netherlands below the age of 50 years eat fewer vegetables than older people. The decline of vegetable consumption may be related to a change of food habits among consumers below the age of 50. They eat more rice and pasta than potatoes. Rice and pasta meals generally contain less vegetables. Convenience may also play a role, after all, since the preparation of vegetables can be time consuming. There are some similarities to be found with drinking of fresh milk, once hailed so much by nutrition and health authorities. Milk consumption declined after the 1960s despite intensive collective promotion campaigns. Vegetables will, however, remain an essential part of the diet, though probably at a lower level than in the 1980s.

Figure 4.1 *Availability of vegetables, fruit and potatoes in the Netherlands percapita day (g), 1938–1990.*

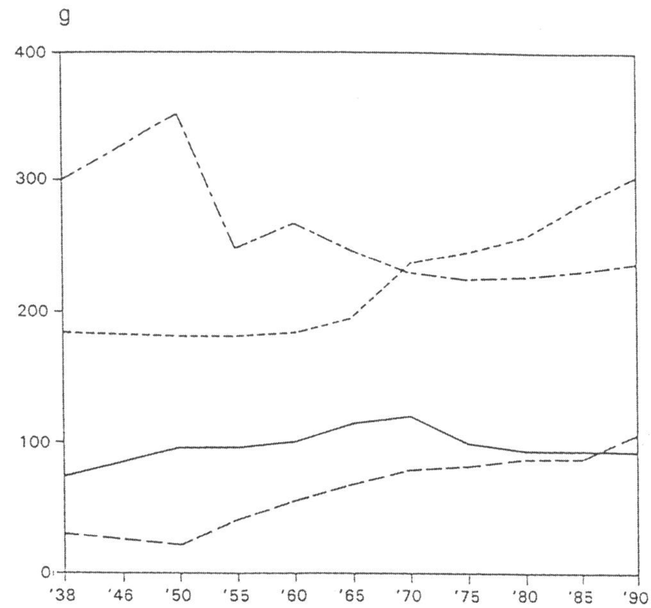

— - — potatoes ---------- vegetables
— — — citrus/ subtropical fruit ________ other fruit

Source: Centraal Bureau voor de Statistiek, Den Haag.

NOTES

1. den Hartog (1998).
2. Burnett (1979), 298–299, 305–306, 311–312; Teuteberg & Wiegelmann (1986), 259–262; Teuteberg 1999).
3. Thouvenot (1987).
4. Burema (1953), 73–75; Mennell (1985), 42–43, 45–46; Jansen-Sieben et al (1989), 26, 33.
5. van Winter (1987), 10.
6. Burema (1953), 73–75.
7. Pitte (1991), 46–48; Slicher van Bath (1987), 292.
8. Burema (1953), 73–75.
9. Giele (1979).
10. Houwaart (1991), 224.
11. *Arbeidsbudgets* (1912).
12. den Hartog, C. & Hautvast & den Hartog, A. (1978), 35; *Nutrition Food and* Health in the Netherlands (1992), 24.
13. Bieleman (1992), 340–343.

14. Jobse-van Putten (1995), 384–411.
15. van Wijk (1985).
16. Bieleman (1992), 71–72.
17. Endhoven (1992), 9–24.
18. Sangers (1963), 76–77, 158–159.
19. Bieleman (1992), 341–342.
20. Sangers (1963), 225–226; Verbraeck (1933), 59–63.
21. Kemmers (1987), 12–13.
22. van Marrewijk (1913), 36.
23. Bieleman (1992), 344–345.
24. *Jaarboek CBT* (1926/1927).
25. *De Tuinderij* 4 (49) (1924).
26. *Jaarboek CBT* (1926/1927), 25–27.
27. *De Tuinderij* 4 (33) (1924).
28. *Jaarboek CBT* (1926/1927), 27.
29. *De Tuinderij* 4 (33) (1924).
30. Ibid.
31. *Jaarboek CBT* (1926/1927), 27–28.
32. *Jaarboek CBT* (1937), 111.
33. *Jaarboek CBT* (1938), 116.
34. *Jaarboek CBT* (1928), 28.
35. *Jaarboek CBT* (1929), 27–28.
36. Ibid., 29.
37. *Jaarboek CBT* (1930), 25.
38. *Jaarboek CBT* (1926/1927), 28.
39. *Jaarboek CBT* (1928), 30.
40. *Jaarboek CBT* (1931), 30.
41. *Jaarboek CBT* (1929), 28; *Jaarboek CBT* (1930), 25.
42. *Jaarboek CBT* (1932), 39–40.
43. *Jaarboek CBT* (1931), 26.
44. *Jaarboek CBT* (1932), 35–37.
45. *Jaarboek CBT* (1936), 24–25.
46. den Hartog (1983).
47. *Jaarboek CBT* (1939), 44–45.
48. *Jaarboek CBT* (1936), 30–31.
49. *Jaarboek CBT* (1937), 37, 136–138.
50. *Jaarboek CBT* (1949), 85–87.
51. *Jaarboek CBT* (1951), 107–110.
52. Schreurs (1991), 57–58.
53. *Jaarverslag Produktschap…* (1965), 89–90.
54. *Jaarverslag Produktschap…* (1962), 133.
55. *Jaarverslag Produktschap…* (1959), 153–154.
56. *Jaarboek CBT* (1970), 25.
57. *Jaarboek CBT* (1966), 24.
58. *Jaarverslag Produktschap…* (1965), 88; *Jaarboek CBT* (1970), 25.
59. Mennell (1985), 89–91; Jobse-van Putten (1987); van Otterloo (1990), 28–154.
60. Leliman-Bosch (1933), 54–56.
61. Baggerman (1983), 27–29.
62. van der Burg (1988), 14–15, 18.
63. Bruinenberg & Winkelman (1985), 53.
64. Voorlichtingsbureau voor de Voeding (1991).
65. *Food and Nutrition policy in the Netherlands* (1983); *Nutrition Food and health in the Netherlands* (1992).
66. Klerk et al (1998).
67. Gerritsen (1995), 6–8.

BIBLIOGRAPHY

Arbeidsbudgets. Jaarbudgets van zeventig arbeidersgezinnen in Nederland [Budgets of labour households], Rapport eener enquête der sociaal-democratische Studie-club, Rapport no. 69 (Amsterdam, 1912).

Baggerman, Matti, *De betekenis van voeding en voedselbereiding in het huishoud- en nijverheidsonderwijs van meisjes vanaf 1880* [The implication of nutrition and food preparation home economics training for girls from 1880] (University of Wageningen MSc thesis, 1983).

Banning, C., *De voeding in Zaandam 1929–1930* (University of Utrecht Ph.D., 1931).

Bieleman, Jan, *Geschiedenis van de Landbouw in Nederland* [History of agriculture in the Netherlands] (C. Meppel, 1992).

Bruinenberg, Ans C. & Winkelman, Margreet L. J., *50 jaar voedingsvoorlichting van de Stichting Huishoudelijke en Consumentenvoorlichting HVP* [50 years nutrition education in rural areas] (Den Haag, 1985).

Burema, L. *De voeding in Nederland van de middeleeuwen tot de twintigste eeuw* [Nutrition in the Netherlands from the middle ages till the twentieth century] (Assen, 1953).

Burg, M. van der, *Een half miljoen boerinnen in de klas. Landbouwhuishoudonderwijs vanaf 1909* [Rural home economics education from 1909] (Heerlen, 1988).

Burnett, John, *Plenty and want. A social history of diet in England from 1815 to present day* (London, 1979).

Endhoven, J. G., 'De boomgaert en warmoesluyden', in: Endhoven, J. G. et al. (eds), *De tuin van Holland, Geschiedenis van tuinbouw en groenteveilingen in Nederland* [History of horticulture auctions in Leiden] (Utrecht, 1992), 9–24.

Food and nutrition policy in the Netherlands, Ministry of Welfare, Health and Cultural Affairs (Den Haag, 1983).

Gerritsen, W. J., 'Naar 2 ons groenten en 2 stuks fruit per dag' [To 200 g of vegetables and 2 pieces of fruit a day]. *Voeding* 56 (9) (1995), 6–8.

Giele, Jacques, *Arbeidersleven in Nederland*, 1850–1940 [Labour life in the Netherlands] (Nijmegen, 1979).

Hartog, Adel P. den, 'North-western Europe and food from the Mediterranean: a century of influence on the Northern diet'. *Revista di Antropologica*, supplemento 76 (1998), 315–326.

Hartog, Adel P. den, 'Food habits in a situation of crisis: the unemployed and their food in the years 1930–1939 in the Netherlands'. *Ernährungs Umschau*, supplement, 30 (1983), 33–38.

Hartog, Cornelis den & Hautvast, Joseph G. A. J. & Hartog, den, Adel P., *Nieuwe voedingsleer* [Handbook of nutrition] (Utrecht, 1978).

Houwaart, E. S., *De hygiënisten, artsen, staat en volksgezondheid in Nederland* [Physicians state and public health in the Netherlands] (University of Maastricht Ph.D., 1991).

Jaarboek Centraal Bureau van de Tuinbouwveilingen in Nederland [CBT, Yearbook of the Central Bureau of Horticulture Auctions in the Netherlands] (Den Haag, years 1926/1927–1940, 1947–1971).

Jaarverslag Productschap voor Groenten en Fruit [Annual Report of the Interprofessional Organization of Vegetables and Fruits] (Den Haag, years 1964–1991).

Jansen-Sieben, Ria & Winter, Johanna M. van, *De keuken van de late middeleeuwen. Een kookboek uit de Lage Landen* [The kitchen of the late middle ages. A cookery book from the Low Countries] (Amsterdam, 1989).

Jobse-van Putten, Jozien, *Eenvoudig maar voedzaam. Cultuurgeschiedenis van de dagelijkse maaltijd in Nederland* [Simple but nutritious. Cultural history of daily meal in the Netherlands] (Nijmegen, 1995).

Jobse-van Putten, Jozien, 'Met nieuwen tijd, komt nieuw (w)eten. Invloed van voedingsonderricht op de Nederlandse voedingsgewoonten' [Influence of Nutrition education on Dutch food habits]. *Volkskundig Bulletin* 13 (1987), 1–29.

Kemmers, W. H., 'De groente- en fruitveilingen tot 1945', in: *100 jaar veilingen in de tuinbouw* [Hundred year of horticulture auctions] ('s Gravenhage, 1987), 12–34.

Klerk, M. et al., *Fruits and vegetables in chronic disease prevention* (Wageningen, 1998).
Leliman-Bosch, M. E., *Geschiedenis van het huishoudonderwijs in Nederland* [History of home economics education in the Netherlands] (Rotterdam, 1933).
Marrewijk, G. J. van, *De Loosduinsche groenteveiling. Haar oprichting, geschiedenis, werking en uitkomsten* [The history of the Loosduinen vegetable auction] (Naaldwijk, 1913).
Mennell, Stephen, *All manners of food. Eating and taste in England and France from the middle ages to the present* (Oxford, 1985).
Netherlands Nutrition Council, *Nutrition, Food and Health in the Netherlands. Summary of the nutrition report 1990* (Den Haag, 1992).
Otterloo, Anneke H. van, *Eten en eetlust in Nederland, 1840–1990* [Eating and appetite in the Netherlands, 1840–1990] (Amsterdam, 1990).
Pitte, Jean-Robert, *Gastronomie française: historie et géographie d'une passion* (Paris, 1991).
Sajet, B. H. & Polak, W., *Eene voedingsenquête in den mobilisatietijd* [A nutrition survey during the mobilization of the First World War] (Amsterdam, 1916).
Sangers, W. J., *Structuur en conjunctuurverschijnselen in de Nederlandse landbouw, in het bijzonder ten aanzien van de groenteteelt* [Structure and cyclical economic movements in Dutch agriculture, in particular to vegetable cultivation] (Wageningen, 1963).
Schreurs, Wilbert, *Collectieve reclame in Nederland* [Collective advertising in the Netherlands] (Leiden, 1991).
Slicher van Bath, Bernard, *De agrarische geschiedenis van West-Europa, 500–1850* [The agrarian history of Western Europe, 500–1850] (Utrecht, 1987).
Teuteberg, Hans-J. & Wiegelmann, G., *Unsere tägliche Kost. Geschichte und regionale Prägung* (Münster, 1986).
Thouvenot, Claude. *Le pain d'autrefois. Chroniques alimentaires d'un monde qui s'en va* (Nancy, 1987).
Tuinderij, De, 'Wat wij met onze reclamecampagne wenschen te bereiken' [What we want to achieve with our advertising campaign]. *De Tuinderij* 4 (33) (1924).
Tuinderij , De, 'Iets uit de geschiedenis van den tomaat' [Something on the history of the tomato]. *De Tuinderij* 4 (33) (1924).
Tuinderij, De, 'Het Reclamefonds' [The Advertising Fund]. *De Tuinderij* 4 (49) (1924).
Tuinderij, De, 'Reclame in het Binnenland' [National advertisement]. *De Tuinderij* 4 (49) (1924).
Verbraeck, A. A., *Het Westland* [The Westland region] (University of Amsterdam Ph.D., 1933).
Voorlichtingsbureau voor de voeding, *Iedere dag uit ieder vak. 50 jaar voorlichtingsbureau voor de voeding* [50 years Bureau of Nutrition Education] (Den Haag, 1991).
Wijk, P. A. M. van (ed.), *Boerderijen bekijken. Historisch boederij-onderzoek in Nederland* [Looking at farm. Historical farm research in the Netherlands] (Amersfoort, 1985).
Winter, Johanna M. van, *Van soeter cokene. Recepten uit de Romeinse en middeleeuwse keuken* [Recipes from Roman and Medieval kitchens] (Bussum, 1987).

MAJA GODINA-GOLIJA

5 Vegetables in the food culture of the Slovenian urban population between 1850 and 1950

In the middle of the nineteenth century Slovenia faced a number of rapid political and economic changes, which greatly affected the everyday life and culture of its people. Because of land reform, which terminated the feudal land tenure system, farmers were no longer obliged to work the land and thus became much more mobile. Nevertheless, their position, worsened, for they were now obliged to compensate their former landlords for the land they had received as well as to pay numerous taxes. This process resulted in land partition and the migration of farmers to seek work away from their native villages. Land reform and the abolition of the feudal relationships between landlords and their serfs caused the breakdown of the feudal system. Farmers became the owners of fields, pastures and vineyards. In 1848 as much as 97.5 per cent of land in the Slovenian part of Austria-Hungary was owned by farmers (75 per cent in Prekmurje). Farmers' rights to own woodland and use manor pastures were partly settled in money, but mostly by the redistribution of forests and pastures. The extent of common land, especially pastures that were distributed between homesteads, increased considerably by the end of the nineteenth century as many farmers moved to towns, obtaining employment in different trades, in manufacturing and emerging industries. The end of the nineteenth century also marked the beginning of emigration overseas to North America and Brazil. Mobility was also facilitated by the completion of the railway between Vienna and Trieste, which connected all the larger towns in Slovenia and which stimulated economic growth, especially in commerce and industry. However, it also destroyed the older forms of transport and their associated trades, such as wagon drivers, river boatmen and their suppliers.

In Slovenia, country towns and provincial centres were quiet, simple places. The original town population was sparse consisting mostly of small tradesmen, craftsmen and shopkeepers, workers in service industries, and domestic servants. Life in these towns was marked by tradition, by provincial narrow-mindedness, patriarchal laws, old-fashioned views, and general poverty, though there were a few noblemen and wealthy townsmen, together with those who worked in the liberal professions, manufacturers, wholesale merchants, higher-ranking soldiers, and politicians. Towns were unattractive, consisting mostly of old, low and simple houses in narrow streets filled with dust in the summer and mud in other seasons. These little towns were more like large villages and did not follow the course of development of towns in western and central Europe. In the last third of the nineteenth century, the first large industrial companies with several thousand employees were opened in Slovenian towns. In 1871 Ljubljana acquired a tobacco factory with over one thousand employees, and in 1863 Maribor a railway machine company, which gave work to more than 1,000 local and foreign workers, was opened.[1]

The period after the First World War, with its abundance of cheap electricity, favourable import taxes and cheap labour migrating to towns from the countryside, resulted in the appearance of numerous factories whose owners were foreign capitalists, and in the subsequent rapid development of industrial mills in Slovenian towns. In addition to people working in various commercial trades and service industries, factory workers became a growing percentage of the town population. In 1931, 27.5 per cent of the population of Ljubljana, the largest town in Slovenia and its capital, worked in industry and trade, while in Maribor, the second largest city, this number rose as high as 30.9 per cent. Many also worked in commerce, banks, and transportation, while other branches of the economy were much less significant.[2]

One important issue that related to urbanization was feeding the growing urban population.[3] Characteristic of the food culture of Slovenian townspeople in the period dealt with in this chapter above all were dishes prepared from vegetables and flour, as well as soups. The following section sheds light on these elements of Slovene food culture.

Characteristics of the Slovene Diet 1850–1950

The food culture of Slovene townspeople depended largely upon the social and financial situation of individuals, but also upon their regional origins. The latter, however, was of lesser importance because of the influence of town residents and of mutual interaction of different social and regional groups.[4] During the century between 1850 and 1950, most inhabitants of Slovene towns ate adequate amounts of rather monotonous food; the same seasonal dishes were often on the menu. This was also due to the fact that many people grew as much as possible in their own gardens or in rented fields, or received produce from their relatives. Food was also the family expense on which they most often saved. By purchasing food of lesser quality many town families, especially working-class families, saved money for other, more visible and high status elements of material culture such as new clothes, house furnishings, or a bicycle.[5] In the 1920s and 1930s the majority of townspeople therefore bought only the most essential staples for their everyday meals: flour, salt, groats, semolina, chicory coffee, sugar, and certain spices. Only occasionally, most often on pay-days, they would buy more expensive, rarely consumed delicacies such as coffee, salami, cheese, fresh meat of good quality, rolls and white bread, sweets, more expensive vegetables, citrus and other fruits.[6]

Prior to 1950 the town population ate mostly dishes made from vegetables, especially legumes; beans were prepared throughout the year, most often shelled and made into soups, sauces and salads, or served as a main dish topped with pork crackling. In the summer, string beans were prepared in soups, sauces and salads, or topped with pork crackling or cream. Peas and lentils were less widely used, as were horse beans, which had been the most widespread legume in Slovenia before being supplanted by the potato. Chickpeas were eaten mostly in western Slovenia, especially in the coastal region where they were prepared in soups and sauces, but also added to bread dough. After the middle of the nineteenth century the potato became the most popular vegetable in Slovenia. The knowledge of how to prepare it spread rapidly, and, prepared in many different ways, it became incorporated into every meal. Cabbage, turnips, and kale were the basic food ingredients in poor town or village families. Fresh cabbage was prepared as a salad; boiled cabbage, turnips or kale were turned into soups and sauces, as well as main dishes, which were enhanced with pork crackling or minced lard. Families in the countryside and in towns alike consumed large quantities of boiled sauerkraut and pickled turnips, which were the most important vegetable dishes in winter.

Apart from bread, the cereal dishes most often found on the table were maize, wheat or buckwheat porridge, and polenta, which in Primorska almost entirely replaced bread at all meals. Popular supper dishes were certain cereal dishes boiled in milk. Millet gruel and different kinds of thin porridge were very popular in the nineteenth century, but were gradually replaced by groats or milk-boiled rice. Gruel and porridge dishes ceased to be consumed in town families before the Second World War.

Meat dishes, which were less important in the food culture of ordinary townspeople, were more often consumed in wealthier families. Soup-boiled beef, pork cutlets or minced meat were often served on an everyday basis, while the choicest cuts, for instance sirloin steak, roast veal, roast pork, roast beef, veal cutlets, capons, and so on, were part of a festive menu. Less affluent households prepared meat only on Sundays or on holidays, serving cheaper meat, for instance beef or smoked pork boiled in water, which was later made into soup. Meat dishes were thus the most important element of meals on festive days, on which even poorer households prepared at least one meat dish. The number of dishes and, above all, the quality, price and quantity of meat, as well as the manner of its preparation all reflected the financial status of individual families.[7]

The importance of growing vegetables in the food culture of the town population

Throughout history, vegetables played an important part in the food culture of Slovenia. The oldest sources for vegetable cultivation are medieval land registers which, among the taxes paid to landlords – the so-called kitchen taxes – also mention the vegetables grown in farmers' gardens.[8] Old surveys and land registers from the eighteenth and nineteenth centuries also mention that vegetables were grown in gardens and fields. During the Middle Ages, vegetables were grown only in the vicinity of consumers because of transportation difficulties, but townspeople also grew them in gardens and in fields on the outskirts, beyond the town walls. Some house owners even had gardens next to their houses within the city walls. Certain guilds also organized their own vegetable gardens outside the town, renting them to their members. The most frequently grown vegetables were the legumes, especially horse beans, lentils, chickpeas and peas, but also cabbage and turnips.

In the period between the sixteenth and the nineteenth century, vegetables still lacked economic value, and were grown merely for individual consumption. Besides the ever-popular legumes, the gardens on the outskirts of towns contained pumpkins, cucumbers, melons, radishes, turnips, cabbage, carrots, onions and garlic. Townspeople in Primorska also grew artichokes and asparagus. Besides vegetables and different kinds of berries, gardens often contained local herbs as well: marjoram, thyme, mint, valerian, basil and parsley.

Social and economic changes in the middle of the nineteenth century affected the development of agriculture as well and thus also the growing of vegetables. This became a profitable business, especially in the vicinity of larger urban centres and in those areas of Slovenia with a mild, Mediterranean climate, which enabled the growing of early spring vegetables. Vegetable growing prospered above all in Krakovo and Trnovo in Ljubljana, and in the vicinity of Gorica, Trieste, Koper, Maribor and Celje. The demand for fresh vegetables increased among townspeople and certain new kinds of vegetables were introduced as well. The gardens of Primorska featured primarily early potatoes, different varieties of lettuce, radicchio, fennel, cabbage, kale, peas, cantaloupes, pumpkins, asparagus and artichokes. Gardens in central Slovenia and Štajerska contained mainly legumes, onions, garlic, various lettuces, potatoes, cabbage, kale, turnips, beets, radishes, carrots, cucumbers, kohlrabi, cauliflower, spinach, and asparagus.[9]

In the first half of the twentieth century, market gardening, – that is, growing vegetables for sale in the vicinity of larger Slovene cities such as Ljubljana, Celje, Maribor, Gorica and Trieste – grew into a professional occupation. Growing vegetables for one's own needs became widespread as well. Less affluent townspeople tried to grow as much as possible in allotments, rented fields or gardens on the outskirts of towns. Each family member participated in this activity and had his or her own allotted chores. It was not unusual for a tenant to receive a garden plot together with an apartment, a practice that enabled one to grow the most essential vegetables for one's daily meals. Those who lived in the town centre where houses did not have gardens had to buy vegetables at stalls in outdoor markets. In affluent town families, which did not have their own gardens, ladies of the house frequented such outdoor markets daily, selecting vegetables and haggling over prices themselves. Their maids then took their purchases home. Before the Second World War the most important market days were Wednesday, Friday and Saturday. Outdoor markets in towns teemed with sellers offering a variety of vegetables from their pushcarts, wagons, tables or even from the ground. With the exception of the potato, which was sold by measuring cups, vegetables were not usually weighed, but sold by the piece or by the heap. Market vendors started to weigh them only in the 1940s.[10]

In addition to those vegetables already meantioned which were popular among Slovenian townspeople in the nineteenth century, the consumption of previously less-known vegetables became widespread during the 1930s: green peppers, zucchini, eggplant, tomatoes, artichokes, green and red radicchio. The changed economic and political circumstances of the time also helped certain kinds of vegetables to spread across Slovenia. Numerous Slovenes, for instance, fled the Italian fascist regime in Primorska and arrived and settled in the Kingdom of Yugoslavia. They brought with them not only elements of their own culture, but also different food habits, for instance, knowledge of how to prepare some new vegetable dishes.[11] Despite these novelties, legumes, especially beans, remained the most important ingredients of everyday meals, as well as potato, cabbage and turnips, onions and certain kinds of lettuce (Table 5.1). This has been further confirmed by data on the sale of vegetables in outdoor markets. In 1936, the Slovenian newspaper reported that 1,377 wagons of potatoes, 165 wagonloads of onions, 286 wagons of cabbage, 7,098 sacks of wheat, 1,291 sacks of beans, 2,836 wagonloads of fruit, and 'mounds' of other vegetables were sold at the Maribor outdoor market.[12] Other urban markets of this kind, for instance the one in Ljubljana, sold approximately the same amount of vegetables.[13]

Table 5.1 *Crops in the Drava Province as a percentage of the cultivated field area in 1939.*

	Districts in Carniola	Districts in Lower Styria	Prekmurje
Peans	0.07	0.06	0.01
Lentils	0.02	0.01	n.a.
Beans	1.04	0.35	0.41
Horse beans	0.02	0.07	n.a.
Potatoes	19.21	19.52	12.71
Turnips	2.59	1.55	0.47
Carrots	1.92	0.54	0.09
Cabbage	1.23	1.12	0.49
Onion	0.19	0.20	0.12
Green peppers	n.a.	0.01	0.06
Tomatoes	0.01	0.04	0.03
Total Vegetables	**26.30**	**23.47**	**14.39**
Cereals	53.90	65.20	76.67
Oil-plants	0.01	0.32	0.12
Soya	n.a.	0.01	n.a.
Poppy	n.a.	0.02	0.50
Flax	0.51	0.17	0.21
Hemp	0.05	0.02	0.17
Hop	0.01	1.37	n.a.
Clover	15.56	7.53	7.35
Without data	3.66	1.89	1.04
Total	**100.00**	**100.00**	**100.00**

Source: Valenèiè (1979), 268.

The most important vegetable dishes

There are numerous ways to prepare vegetables. As well as boiling in water, which is most commonly used, they can be braised, sautéed, baked, or fried. Between 1850 and 1950 urban as well as rural households usually boiled vegetables in water, only rarely preparing them in a different manner. Served as the main dish, boiled vegetables were topped with minced lard or pork crackling, on fast days also with butter; more prosperous families used butter, fried bread crumbs, mayonnaise or cream on more expensive vegetables. The very first cookery book in Slovenian, written in 1868, was *Slovenska kuharica ali navod okusno kuhati navadna in imenitna jedila* (Slovene Cookery Book or How to Prepare Ordinary and Fancy Dishes Tastefully). Its author, Magdalena Pleiweis, listed as many as 36 recipes for the preparation of kohlrabi, turnips, cabbage, carrots, kale, peas, beans, radishes, and also more expensive and rare vegetables such as cauliflower, spinach, asparagus, artichokes and zucchini.[14] The number of recipes for the preparation of vegetables increased steadily in cookery books from the first half of the twentieth century, and some new ones, for instance for the preparation of sautéed, fried and stuffed eggplant, were added.[15] The latter were especially popular among the housewives of more prosperous urban families. The majority of the urban population, however, stuck to the familiar, well-tested techniques for decades. Shelled beans went into soups and gravies, or were boiled and topped with pork crackling. This was a typical dish on fast days, consumed in most families regardless of their social background. Restaurants offered it as well, especially on Fridays. String beans were very popular in the summer, prepared in soups and gravies, or as a salad, boiled and topped by fried bread crumbs or cream. Lentils were a typical winter dish and were cooked in gravy or prepared as a salad. Only very poor households still ate horse beans at the beginning of the twentieth century. Peas, on the other hand, were considered a fancier vegetable and were therefore consumed mostly by the prosperous, who prepared them in gravy, stewed with carrots, or boiled in the pods and dressed with lard.

Cabbage, turnips and kale were important, often-served staples in poorer families. In winter boiled pickled turnips or sauerkraut were served with mush, beans or potatoes. In the summer these households often served stewed sweet cabbage, cabbage soup or pasta with cabbage and a salad, or stewed kale gravy. More prosperous families prepared cabbage, turnips or kale only as a side dish to meat or to beans on a fast day, or as a salad.

Potato dishes were popular in different regions and among different social classes. After the end of the nineteenth century, when the potato became a familiar and often-served vegetable among all social classes, its importance in the food culture of Slovenia increased.[16] It was made into soups; as a side dish it was popular in the form of chipped or baked potatoes, it was boiled and topped with lard, mashed, or in more prosperous families served as French fries by the beginning of the twentieth century. Cooks prepared potato gravies, mashed potatoes, boiled or fried potato dumplings, and *gnocchi*. Potato salad, considered a fancy dish in the nineteenth century, now became a popular everyday dish. Well-to-do households prepared the salad with mayonnaise and spices, poorer families with onions, oil and vinegar.[17]

The more expensive spinach, cauliflower and asparagus were served in the spring, mainly in more prosperous households and in restaurants, but not as often as the above-mentioned vegetables. Unlike nowadays, outdoor markets were literally flooded with asparagus, which represented typical spring fare for the more affluent. Asparagus was served either baked with sour cream, or boiled and topped with fried breadcrumbs.

Very few urban families served tomatoes, green peppers, zucchini or artichokes. Of these, tomatoes, brought from the coastal regions by immigrants from Primorska, were most frequently consumed. Following the example of gardens near the coast, townspeople elsewhere started to plant tomatoes in their gardens, using them for the preparation of a variety of dishes, mostly as a supplement to other dishes such as soups or gravies. Tomatoes were not eaten as an independent dish, for instance in a salad. Before the Second World War they were mostly made into tomato sauce which was added to soups and gravies in the winter.

Lettuce was eaten in season and grown mostly in local gardens. Popular spring salad vegetables included dandelion, lamb's lettuce, early non-hearting lettuce, or the local Ljubljana iceberg lettuce. Green and red chicory, which had come from the coast with immigrants, became popular at the end of the 1920s. Salads were made from other vegetables as well: beans, lentils, cucumbers, cabbage, potatoes. In winter, when lettuce was very expensive and could therefore be purchased only by the rich, less affluent families prepared turnip tuber shoots, which could be obtained at open-air markets.

Conclusion

Throughout history the role of vegetables has been extremely important in the food culture of Slovenia. In the time period dealt with in this chapter, 1850–1950, despite numerous economic changes, vegetables, together with cereal dishes, remained the most important food component of the diet for both urban and rural population. The consumption of vegetables meant that many less affluent families were able to grow most of the ingredients for their daily meals without any expensive tools or large areas of land. Vegetables were not physically difficult to grow and all family members, including young children, could help with this work. People generally grew the less demanding vegetables, which were sure to yield a crop every year: potatoes, beans, cabbage, turnips, kale, onions and various kinds of lettuce. Vegetables were thus ideal for home growing on small patches of land.[18] Domestic gardens produced the vegetable ingredients for a number of dishes, most often the differently prepared legumes and tuberous plants consumed during main meals such as lunch or supper, or light meals in between.[19]

Vegetable dishes played an important role in the food culture of Slovene farmers, too. Before the Second World War farmers were the predominant segment of the population. Vegetable dishes were part of the everyday menu since some meat and farinaceous foods were almost unattainable for the majority of Slovenes. In their fields farmers usually grew cabbage, turnips, squash, carrots, kale, horse beans, lentils, beans and peas. These vegetables were prepared for any of the daily meals. A popular breakfast dish consisted of either cabbage or turnips with maize, wheat or buckwheat porridge, potatoes or horse beans. Some families also ate potato soup and jacket potatoes for breakfast. Lunch or dinner often consisted of jacket potatoes, sauerkraut or pickled turnips with lard, potatoes with lettuce, squash with flour fried in lard and served with potatoes, turnips with millet or buckwheat porridge, or beans or horse beans with lard and with fresh cabbage or sauerkraut.

During long, snowy winters, when vegetables were scarce, Slovenes ate mostly sauerkraut or pickled turnips which were served on a daily basis, sometimes more than once a day. Cabbage water, a rich source of vitamin C, was also popular. Instead of salad greens, which were difficult to obtain and expensive, townspeople and farmers alike also ate turnip tuber shoots. For winter use some vegetables, for instance potatoes, yellow carrots, turnips or squash, and later also chicory, were kept in clamps, pits which were up to a half a metre deep. Such vegetables were first covered with hay, then with soil; in some places clamps had a wooden lid, additionally protected by a layer of soil for better insulation.

In winter, urban housekeepers from wealthier families bought their vegetables in open-air markets or from greengrocers who had imported them mostly from the vicinity of Trieste or Gorizia. Because of their mild climate these areas had flourishing market gardens. As early as January or February, greengrocers were thus able to offer young onions, early potatoes, chicory, lettuce and lamb's lettuce, dandelion, peas, spinach and artichokes. Less affluent urban households, which could not afford such expensive vegetables, consumed them only rarely; instead they ate dishes made from sauerkraut, pickled turnips or legumes, whose taste had been improved with onions, garlic or lard. Such winter meals were modest and monotonous, and many times there was also a shortage of food.

In times of want or financial difficulties vegetable dishes replaced the traditional breakfast fare of Slovene urban families as well. Salads or vegetables thus replaced the popular café-au-lait prepared from milk and coffee substitutes, which was ordinarily consumed with bread or mush; money required for milk could thus be used for something else. The expensive vegetables were harder to grow and also much more difficult to prepare for consumption. These rarer vegetables, which on occasions replaced simple boiled vegetable dishes, were artichokes, eggplant and cauliflower, either fried and stuffed, or like asparagus, served with different sauces. These dishes gradually spread from the richest households to the less affluent ones and to better restaurants until they became popular among numerous urban households in the 1970s and 1980s. Such vegetable dishes, once consumed mostly during holidays or on Sundays, became common. As with other elements of the food culture of the urban population, regional differences did not play a very important role in this process. More decisive were social differences which affected the way dishes were made or consumed, and how frequently. Only the wealthy could afford expensive, often imported vegetables, which were prepared according to foreign, complicated recipes, on a daily basis. For the majority of the urban population, such vegetables remained inaccessible for most of the twentieth century.

NOTES

1. Godina (1992), 17; Kremenšek (1980), 24.
2. Zagar (1994), 20; Godina (1992), 27.
3. For an analysis of the influence of urbanization on food culture see for instance Braudel (1988); Kisban (1986); Teuteberg (1972).
4. Godina-Golija (1996), 150. Immigrants from Primorje as well as Serbian soldiers contributed elements to Mediterranean and Serbian cuisines. During the time period encompassed in this chapter, these elements already started to blend, though to a smaller extent at first, with the traditional Austrian cuisine, which had been characteristic for the majority of the urban population during the period of the Austrian Empire. The food culture of townspeople also changed due to numerous innovations in the food industry which were the result of intensive developments in the nineteenth and at the beginning of the twentieth century.
5. Godina (1990), 33.
6. Godina-Golija (1996), 39.
7. Godina-Golija (1996), 82. For the relationship between vegetable and meat dishes in general see Harris (1985).
8. Valenèiè (1979), 273.
9. Ibid., 277.
10. Godina-Golija (1996), 43.
11. Ibid., 72.
12. Ibid., 45.
13. Zidov (1994), 112.
14. Pleiweis (1994), 73–86.
15. *Za vsak dan* (s.a.), 500.
16. Godina-Golija (1998), 29.
17. Wiegelmann (1992), 85.
18. Ziessow (1992), 40.
19. Wiegelmann & Mauss (1986), 92.

BIBLIOGRAPHY

Braudel, Fernand, *Strukture vsakdanjega Zivljenja: mogoèe in nemogoèe* [The Structure of Everyday life: The Limits of the Possible] (Ljubljana,1988).

Godina, Maja, *Iz mariborskih predmestij* [From Maribor's Suburbs] (Maribor, 1992).

Godina-Golija, Maja, 'Pomen uvajanja krompirja za spreminjanje prehrane na Slovenskem' [Changes in Slovene Food Culture as a Result of the Introduction of the Potato]. *Traditiones* 27 (1998), 25–38.

Godina-Golija, Maja, *Prehrana v Mariboru v dvajsetih in tridesetih letih 20. stoletja* [Food Culture in Maribor during the twenties and thirties of the twentieth century] (Maribor,1996).

Harris, Marvin, *Good to eat: Riddles of Food and Culture* (New York, 1985).

Kisbán, Eszter, 'Food habits in change. The example of Europe', in: *Food in Change. Eating Habits from the Middle Ages to the Present Day* (Edinburgh, 1986), 3–19.

Kremenšek, Slavko, *Etnološko prouèevanje Ljubljane novejše dobe* [Ethnological study

of Ljubljana's new era] (Ljubljana, 1980).
Pleiweis, Magdalena, *Slovenska kuharica* [Slovenian cookery book] (Ljubljana, 1994, 1st edn 1868).
Teuteberg, Hans J. & Wiegelmann, Günter, *Der Wandel der Nahrungsgewohnheiten unter dem Einfluss der Industrialisierung* (Göttingen, 1972).
Valenèiè, Vlado, 'Vrtnarstvo' [Gardening], in: *Zgodovina agrarnih panog* [History of agriculture] (Ljubljana, 1970), 274–279.
Wiegelmann, Günter, 'Kartoffelspeisen des 19. Jahrhunderts in Norddeutschland', in: *Geschichte und Zukunft einer Kulturpflanze* (Cloppenburg, 1992), 79–89.
Wiegelmann, Günter & Mauss, Annette, 'Fischversorgung und Fischspeisen im 19. und 20. Jahrhundert', in: H. J. Teuteberg & G. Wiegelmann (eds), *Unsere tägliche Kost. Studien zur Geschichte des Alltags* (Münster, 1986), 75–93.
Zagar, Janja, *Oblaèilna kultura delavcev v Ljubljani med prvo in drugo svetovno vojno* [The dress culture of workers in Ljubljana between the First and Second World Wars] (Ljubljana, 1994).
Za vsak dan [For every day], Andrej Pceljnikov (ed.), (Maribor, s.a. [before the Second World War]).
Zidov, Nena, *Ljubljanski Zivilski trg* (Ljubljana market place] (Ljubljana, 1994).
Ziessow, Karl, Heniz, 'Kolumbu's spätes Erbe: Dreihundert Jahre Entdeckungsreisen der Kartoffel in Europa', in: *Geschichte und Zukunft einer Kulturpflanze* (Cloppenburg, 1992), 15–44.

BEAT KÜMIN

6 Eat in or take away

Food and drink in Central European public houses around 1800

A notable trend in recent social and economic history has been the increasing interest in aspects of consumption and distribution. In addition to looking at methods and levels of production, scholars now investigate in more detail how consumers chose and obtained the various goods they needed to buy. A survey of these studies suggests that the roots of modern consumer society reach far into the Ancien Régime.[1] The revolutions of the 'long' nineteenth century brought a dramatic expansion of economic activities and fundamental reorganizations in the fields of transport and trade,[2] but this should not obscure the fact that an extensive system of goods exchange was in place well before 1800 and that a basic commercial infrastructure had emerged even in many villages.[3]

In food history, too, institutional, social and cultural dimensions of marketing and consumption have received growing attention.[4] One significant aspect, however, the provision of food and drink in public houses, has as yet been largely overlooked.[5] Some important insights have emerged in the field of alcohol studies,[6] but on the whole inns, taverns and alehouses are still primarily studied as centres of sociability, feasting and popular culture.[7] It will be argued in this chapter that commercial hospitality was an important part of the 'landscape of food' around 1800, but that its role varied in accordance with location, context and clientele.[8] The evidence derives from official registers of public houses, administrative sources and a variety of local records, but also contemporary descriptions and travellers' reports. In order to obtain information from as broad a background as possible, two very heterogeneous case studies have been selected: the Swiss City Republic of Bern and the Electoral Princedom of Bavaria. These provide insights into a very large territory (containing nearly 1.5 million people in the late eighteenth century), two different religious denominations (Zwinglian and Catholic) and two distinct political regimes – a strong communal tradition in Bern as opposed to central princely government in Bavaria.[9] The discussion which follows sketches the circumstances in which public houses offered their services to consumers. Part two will consider questions of production and distribution, examining how innkeepers obtained the drinks and foodstuffs they sold in their establishments.

Eating and drinking in early modern inns

There were two basic types of commercial hospitality at this time:
1) Taverns and alehouses, restricted to the provision of particular drinks, bread and cheese;
2) Inns, with much more extensive rights, including the sale of a broad selection of alcoholic beverages and cooked meals, licensed to cater for large parties of guests and the ability to offer overnight accommodation.[10]

Taken together, official registers contain over 1,000 such establishments in the Republic of Bern at the end of the eighteenth century, i.e. one per 394 inhabitants overall, but with much regional and local variation. To take an extreme case, in some Bavarian pilgrimage sites, the density of inns could be as high as one per 25 inhabitants.[11] As the following discussion will be concerned with food as well as drink, the emphasis will be on inns with their extensive catering rights. Even so, we are looking at a large number of institutions. In the French-speaking part of the old Republic of Bern, for example, contemporaries recorded 1 inn per 670 inhabitants in 1786, compared to one hotel per 1,500 heads of population according to the statistical count for 1996.[12]

Commercial hospitality could be found wherever people were likely to gather in sufficient numbers. Towns and cities with their many economic and administrative functions were obviously well-provided for, but other 'centres' such as sites of religious institutions, courts, customs points, river crossings and important transit routes were equally prominent. The variety of locations was reflected in a wide range of different customers. Some inns accommodated representatives from all social groups, others were more specialised, catering particularly for local people, merchants, journeymen, carriers, pilgrims or early tourists. This heterogeneous clientele had very distinctive needs and innkeepers provided increasingly tailor-made services. Five typical situations for public consumption can be distinguished.

First, of course, many patrons simply wanted a drink. Being used to water or milk in everyday life, rural people drank alcohol mostly on convivial occasions,[13] and there is no shortage of contemporary witnesses complaining about frequent excess. In his description of the district of Laupen in 1779, the Bernese law student, Niklaus Holzer, remarked that "moderation was not one of the main characteristics of this people [...] in drinking even less than in eating. The use of wine is omnipresent, its abuse particularly strong among the inhabitants on this side of the river Aare. Most of the wine is ruined by fraud and mixing, it is rarely very alcoholic, but has a strong taste of sulphur. It causes headaches and dullness. In order to get rid of this, people are forced to return to the tavern the day after, sometimes for weeks on end".[14] The quality of local wine, as Holzer emphasized, was often terrible: the Bernese government heard of an alarming case in the village of Gümligen in 1786, where the innkeeper was accused of offering "such bad wine [...] that one had to fear for the health of his guests".[15]

Nevertheless, tax records reveal large quantities of drink consumed in public houses. In 1788, the keeper of the *Sun* inn at Herzogenbuchsee, a place little more than a large village on an important transit route where three fairs were held, accounted for no less than 38,700 litres of wine sold over the course of one year, equivalent to a turnover of some 106 litres a day. The most exclusive of the nine inns in the capital Bern, the *Falcon*, paid dues on a total of 18,000 litres over the same twelve-month period and here the quality must have been rather more agreeable, as guests could choose from an extensive drinks menu which included sweet dessert wine and even Champagne.[16] In the district of Dachau near the capital of Bavaria, meanwhile, inns and alehouses accounted on average for some 4,146 litres of brown beer sold to their customers in the tax year 1788–1789, but some establishments did considerably better: the innkeeper at Schweinbach recorded a figure of no less than 22,303 litres and the highest volume in the entire Princedom was reached by Magdalene Reifenstallin, landlady at Gmund on the Tegernsee, with a remarkable 74,300 litres over the whole year or 203 litres per day.[17] Brandy was also sold, but apparently not in excessive amounts.[18]

Second, inns provided meals for travellers of all kinds. The building of new roads and the establishment of more frequent and affordable coach services had boosted the numbers of wealthy tourists as well as visitors from humbler backgrounds. This increase in spatial mobility depended on an adequate network of inns.[19] Previously, public ordinances had specified very rich meals to be served to all guests at a fixed price,[20] but by the eighteenth century there was more variety in quality and quantity. Many people asked for smaller menus, as they were neither used to eating much meat nor able to afford it.[21] Catholic areas like Bavaria also insisted on the observance of fast days, even though visitors of other denominations were granted an exemption in 1796, as long as they ate meat in their private rooms.[22] On top of central rules there were local guidelines, as in a lease issued in 1759 for the inn at Spötting near Landsberg, which instructed the innkeeper not to offer his guests any "precious dishes such as game, birds or fish".[23]

Travellers' reports suggest a wide variety of standards encountered in public houses. Some innkeepers were unable to produce more than sour wine, stale bread and mouldy cheese, but in the small Bernese town of Wiedlisbach in 1773, the supper menu included trout, roast chicken and strawberries, accompanied by a bottle of good wine from the Neuchâtel region.[24] According to a contemporary travel guide, Swiss inns usually offered a soup, three main courses and a dessert for one florin, which was not cheap.[25] A leading house such as the *Falcon* in Bern, however, satisfied even the finest palates, catering for state guests, visiting princes and members of the local *haute volée* with equal aplomb. Its menu book, a rare eighteenth-century survival, records the large number of important customers, the large sums of money invested in early modern haute cuisine and the availability of out-house catering.[26]

Third, inns were entitled to hold banquets for large groups of people. Professional organisations without their own guild halls relied on commercial hospitality to host their functions and private people did the same when it came to the celebration of rites of passage. In Bavaria, in fact, innkeepers enjoyed a legal monopoly over engagements, weddings and funeral meals.[27]

Ordinances at Ebersberg made it explicit that they had the "exclusive right" to cater for such occasions and that taverners or alehouse keepers were not allowed to do so. Many lords even forced their subjects to hold engagements and weddings only in the local manorial inn. At Neuburg near Passau, the innkeeper possessed "a lease dating from 1 March 1781, which included the subjects' obligation to hold weddings and large banquets in his house, or else to deliver half of the cost to the lord's chancellery, which had to pass the sum on to the innkeeper". By the early nineteenth century, however, such restrictions had become illegal.[28] It goes without saying that an event of this kind involved enormous quantities of food and drink, as both rich and poor people considered it a matter of honour to hold lavish weddings.

Fourth, commercial hospitality was needed to cope with the large influx of people on extraordinary occasions. In towns and other commercial centres, market days were particularly busy and called for additional efforts in the catering trade. An agreement between butchers and innkeepers in the district of Aarwangen near Bern in 1684, which allowed the latter to slaughter one piece of cattle for every party of seventy or more guests during local markets, points to the large number of people who had to be looked after by individual establishments. Some places even granted wine or beer taverns temporary licences to serve meat during the local fairs (two examples being the *Deer* at Langnau in the Emmental and the alehouse at Aholming in Lower Bavaria) or they established further ad-hoc catering institutions to cope with exceptional demand, allowing bakers, members of other trades and ordinary burghers to sell food and drink to the many visitors. At Kelheim in Bavaria, one inn was "exercised only during market time" and owned by Mathias Vogl, who normally plied his trade as a saddler. In the small city of Laupen near Bern five additional people – among which the butcher Johannes Schlatter – were licensed to run a tavern during the spring and autumn fair of 1753, and some time later they accounted for no less than 2,200 litres of wine sold on these occasions, about a quarter of what the ordinary *Bear* inn used over a whole year![29]

Fifth and finally, a point which is often overlooked, inns provided local people with ready-made meals as well as drinks and uncooked foods to take away. In 1770, the inhabitants of the village of Neuenegg near Bern petitioned their government for the establishment of a permanent butcher, as so far "they had been forced to either fetch their meat from distant places or to buy it ready-prepared from the local inn", both of which proved costly and cumbersome to them.[30] In an official register compiled in 1786, about half of all Bernese inns exercised the right to slaughter cattle and/or to bake bread, some for domestic usage only, but others with the explicit permission to sell to members of the public. In the village of Oberhofen on the Lake of Thun, it was said that the innkeeper had always enjoyed baking rights, "because there was no baker in the village, and that he could sell the bread to travellers as well as locals".[31] Innkeepers clearly appreciated these privileges as an important part of their livelihood. When a butcher asked the Bernese council to be allowed to exercise his trade in eighteenth-century Gümmenen, an important river crossing point on the road between Bern and Murten, arguing that the two local innkeepers were not properly qualified and unable to meet existing demand, the accused put up fierce resistance, imploring the authorities that they would face economic ruin should they no longer be allowed to sell meat to the public.[32]

Hundreds of their Bavarian colleagues exercised similar rights. Examples for the general sale of meat include the inns at Lederdorn or Strasskirchen, while the regional authorities at Wolfratshausen repeatedly asked for similar permission to be granted to the alehouse keeper at Arget, where no butcher was to be found. *Jesenwang* can serve as an example of an establishment allowed to slaughter for the needs of its guests only. Bread could be purchased e.g. from the landlords at Litzeldorf or Rothenberg, and at Bernried, it was specified that the inn's baking rights had to be exercised by a master baker. In the manor of Neuenfrauenhofen, the lord had licensed eight alehouse keepers to sell bread to the local population with the telling explanation that "the subjects lived too far away from existing breweries and inns". At Neuherberg just north of Munich, the innkeeper sold on average bread worth one florin a week in 1754, a substantial sum, considering that his servant received an annual wage of 20 florin.[33]

People also used inns and taverns if they needed to buy drinks for private functions. The taverner at Vallorbe in the French speaking part of Bern was one of many permitted to sell wine "in pots and pints" for consumption off the premises, the inn at Triftlfing in Bavaria existed for the "convenience" of the locals, because the local lord's brewery did not cater for individual customers, and the public house at Niederviehbach seemed to be essential to relieve the inhabitants of nearby Gummering, who used to have to fetch their beer from "distant places".[34] Many inns even acted as early modern versions of mini-markets or drugstores, with countless Bavarian innkeepers allowed to sell brandy, salt, tobacco and a whole range of other everyday goods. Around 1800, such service centres existed for instance at Stephansposching, where the innkeeper Anton Tuschl also ran a grocery, butchery and bakery, while his colleague at Hepberg exercised a "grocery, because there was no other grocer around", even though he possessed "no formal licence" to do this at all.[35]

Providing for early modern inns

There can thus be little doubt that a large amount of food and drink was available in public houses. But where did these provisions come from? It has been shown that substantial numbers of innkeepers exercised baking and slaughtering rights, and in turn there were master butchers and bakers who obtained an inn to augment their income.[36] If no such combination existed, owners had to make purchases and here one sometimes relied on another. The innkeeper of Hermiswil in the Emmental, for instance, regularly visited cattle markets in the area, where he bought meat for himself and colleagues at Burgdorf and Hegen.[37] Trading activities were a general feature of the profession and innkeepers seem to have acted as brokers for nearly every imaginable transaction. In the country, furthermore, peasants often turned to the hospitality trade to earn some extra income, like the alehouse keeper at Kasing, who was described as being involved in "field economy". Many innkeepers held substantial pieces of land (at Wiesing it was a whole *Hof* or farmstead), meadows, and gardens by means of feudal tenure, others were wealthy enough to buy property successively, as and when the opportunity arose. Wolfgang Lechner, innkeeper at Obermenzing in 1715, owned several plots of land within and without the village borders, nine horses, 14 cows, one bull, two calves, 19 sheep and three pigs. Again, one activity benefited the other: agricultural surpluses could be sold in the form of meals to visitors, augmenting the economic prospects of an innkeeper.[38]

As for the provision of alcohol, it mattered which kind of drink dominated in the area. In the late eighteenth century, Bavarian customers drank predominantly beer, Bernese patrons mainly wine. As more or less anywhere, keepers of public houses in the Swiss city republic simply sold their own produce if they happened to be wine-growers themselves – at Grens in the Vaud it was described as a "natural law".[39] Those without their own vineyards were obliged to buy from the patricians who made up the government and monopolised the trade. The latter also supplied some 200 cellar taverns in their own houses along the streets of the capital, and taken together, all innkeepers, taverners and traders of the city of Bern sold 27,000 hectolitres of wine in 1788, an astonishing 228 litres per head of the population![40] In Bavaria, meanwhile, brewing was tightly regulated by the state and local lords: beer could only be produced in licensed breweries, of which there were 337 in 1580 and several thousand in 1847.[41] Many keepers of public houses, however, also owned breweries, allowing them to benefit from similar advantages as those who produced their own meat and bread. At Pfatter, no fewer than seven out of eight public house keepers were described as licensed brewers in 1806 and at Markt Bogen all of the twelve inns belonged to local brewers, who were also allowed to distil and serve brandy.[42]

Conclusion

For the period around 1800 there are no sophisticated consumer surveys to determine where exactly people obtained their daily necessities.[43] The role of commercial hospitality in food and drink supply thus has to be examined indirectly, through the various circumstances in which people used public catering.

The range and quantity of victuals sold in public houses depended on the type and location of an establishment as well as on the specific situation. Alehouses and taverns were normally restricted to the provision of alcoholic beverages, while inns could offer a full catering service. By 1800 a dense network had appeared in town and country, wherever there were enough people to ensure economic viability. All inns, of course, counted both locals and strangers among their customers, but the main emphasis in terms of food and drink supply varied in accordance with the location: in towns and other commercial centres, inns primarily catered for the large number of strangers flooding in on personal business or on special occasions such as markets and religious festivals. A foreign clientele of merchants and travellers also dominated in public houses located on the main transit routes. In smaller, more isolated rural settlements with fewer grocery shops and victualling trades, however, inns proved particularly important for the local population, offering meat, bread and other commodities to take away.[44] The 'market share' of public houses in overall food and drink consumption should not be underestimated: it must have been particularly high for alcoholic beverages and meat dishes, both of which were rare in most people's diets. Furthermore, while many contemporaries visited inns only irregularly, they did so on occasions like weddings or feast days when they ate and drank much more than usual.

A closer look at how innkeepers obtained their victuals reveals a remarkable degree of self-sufficiency. Many were engaged in at least one other profession, be it baking, brewing, some sort of trade or farming, allowing them to sell a considerable proportion of home-made products. The advantages were obvious: synergies for the innkeeper, lower costs for both owner and customers and a higher likelihood of fresh and uncontaminated food. Those things that needed to be bought boosted the victualling trade between town and country. Bern's rural keepers purchased wine and luxury supplies from the capital, while their urban colleagues relied on basic foodstuffs delivered from the country. Intensive exchange between burghers and peasants, in fact, would have been unthinkable without an adequate network of inns and taverns to cater for the merchants and small traders linking the two.

What difference did the religious and constitutional framework make? Judging from the limited basis of our two case studies, the religious denomination mattered in some respects. Fasting rules and the pilgrimage trade were restricted to Catholic areas, moral regulations such as bans on certain entertainments feature more prominently in Zwinglian Bern. Bavaria, a more stratified society with a strong nobility, paid more attention to differences in social status, specifying different rules for lords and humbler subjects.[45] And yet, the basic structures and features of the hospitality trade in Bern and Bavaria look much alike: inns and taverns constituted the basic types, there were comparable efforts of state regulation and similar problems of enforcement. Princes as well as city councils carefully legislated on numbers, prices and standards, and both never stopped complaining what little effect their edicts had.[46]

Around 1800, to close with a view to later developments, inns were truly multifunctional institutions. On top of their undisputed place in social, political and cultural life, they played a significant part in the supply of victuals and other economic activities.[47] This combination of different functions and services in one single place did not survive into the modern period: a gradual process of differentiation, which intensified during the nineteenth century, resulted in a fragmentation into restaurants, shops, clubs, theatres, court rooms, dance halls and many other specialized institutions. While being rediscovered as an object of historical investigation, the 'old inn' has almost fully disappeared from the European landscape.[48]

NOTES

Abbreviations:
Bay HstA: Bayerisches Hauptstaatsarchiv, München
StA BE: Staatsarchiv Bern

1. Sandgruber (1982); Tanner et al. (1998).
2. See most recently Spiekermann (1999).
3. Radeff (1996), esp. 429–431; Kümin & Radeff (2000).
4. Scola (1992); Teuteberg (1997).
5. Cf. the verdict in Schärer (1992); a social and anthropological approach to consumption in public houses around 1500 is offered by Hürlimann (1998).
6. Clark (1985); Hirschfelder (1994).
7. Potthoff & Kossenhaschen (1933/1996).
8. The essay is based on research for a larger project on the 'Social History of Early Modern Inns and Taverns', supported by the Swiss National Science Foundation.
9. Introductory surveys on these areas are provided by Meyer (1981) and Schuck (1999).
10. For the legal and institutional framework see esp. Zeiss (1927) and Stürler (1975).
11. StA BE, B V 148, passim (Bern); Hersche (1997), esp. 326 (Bavarian pilgrimage sites).
12. For calculations and context cf. Kümin (1999).
13. Sandgruber (1982), 181; alcohol consumption was more common in towns: ibid., 182; Hirschfelder (1994), 118.
14. Holzer (1984), 44.
15. StA BE, B VIII 392, p.152.
16. StA BE, B VIII 517, passim (Falcon), and ibid., B VIII 616, p. 6 (Herzogenbuchsee).
17. Bay HStA, GR Fasz. 1551, Nr. 2, part 3 (incl. a volume containing a list of beer sales 1788–1789, fols. 8v–10v, 30v).
18. A mere 25 measures were used at Obermenzing in 1787. Thurner (1986), 270.
19. Behringer (1999), esp. 81.
20. According to the 'Landts– vnd Policeÿ Ordnung der Fürstenthumben Obern vnd Nidern Bayrn' of 1616, 3. Buch, 3. Tit., 1, keepers should serve two different wines and no more than six courses, including two varieties of cooked meat or two courses of fish dishes (one fried and one boiled), with cheese or fruit served at the end. Art.
21. Sandgruber (1982), 157.
22. Meyr (1784–1799), v. 511 f. (1796).
23. Bay HStA, GR Fasz. 872, Nr. 175 ('Freistiftsbrief' of 1759).
24. Gercken (1784), ii. 206.
25. Ebel (1793), i. 22. The same amount of money bought ten pounds of white bread.
26. Extracts edited in Rageth-Fritz (1987).
27. Zeiss (1927), 210.
28. Bay HStA, GR Fasz. 878, Nr. 186, pp. 194 (Ebersberg), 757 (Neuburg).
29. Ibid., Nr. 186, pp. 640 (Aholming), 499 (Kelheim); StA BE, A V 1114, f. 80v (Laupen); for a detailed analysis of the relationship between markets, fairs and the hospitality trade, including the situation at Aarwangen and Langnau, cf. Kümin and Radeff (2000).
30. StA BE, A V 1113, p. 795 f.
31. StA BE, B V 147, p. 326.
32. StA BE, A V 1113, pp. 733–740 (1766).
33. Bay HStA, GR Fasz. 878, Nr. 186, pp. 214, 321, 700, 770, 75, 562, 565, 614; ibid., HR I 235/149 (2), 327v (Neuherberg).
34. StA BE, B V 147, p. 1152 (Vallorbe); Bay HStA, GR Fasz., 878, Nr. 186, pp. 188 (Triftlfing), 729 (Niederviehbach).
35. Bay HStA, GR Fasz. 878, Nr. 186, pp. 623 (Stephansposching), 93 (Hepberg); dozens of examples for inns with small shops attached to them ibid., passim.
36. In 1806 two 'Fleischhauer' (butchers Michael Huber and Paul Heininger) ran inns at

Obernzell (ibid., GR Fasz. 878, Nr. 186, p. 463), two bakers did the same in Breitenberg (ibid., p. 472).

37. StA BE, DQ 478.2, pp. 10–14 (1782).
38. Bay HStA, GR Fasz. 878, Nr. 186, pp. 97 (Kasing), 598 (Wiesing); ibid., GL 653/282 (Obermenzing); early modern estate records reveal many cases of extensive landholding by innkeepers (e.g. the £2 payable for the inn, garden and lands belonging to the innkeeper of Roggwil in the Republic of Bern: StA BE, B V 147, pp. 63–4; 1682).
39. StA BE, B V 147, 1005.
40. StA BE, B VIII 517, p. 60; for legislation governing the wine and hospitality trade in early modern Bern cf. Rennefahrt (1966), 198–255.
41. Figures from Bay HStA, StV 1853, passim, and Drexler (1997), 30.
42. Bay HStA, GR Fasz. 878, Nr. 186, pp. 360–361 (Pfatter), 573–574 (Bogen).
43. For a present-day study see Bopp (1991), esp. chapters 3 (consumer behaviour) and 7 (statistical analysis of food providers).
44. Towards the end of the Ancien Régime, the number of small tenants or landless people who needed to buy daily provisions was growing. For survey works on the economic development in the two case studies see Schremmer (1970) and Radeff (1996).
45. According to the 'Landts– vndAriceÿ Ordnung' (1616), servants could only be served food and drink approved by their masters.
46. Elaborated – for Bern – in Kümin (1999).
47. For medieval origins see Kerntke (1987); for early modern western Switzerland see Radeff (1996), 223–224, for evidence for wholesale trading of foodstuffs in public houses see Scola (1992), 152, 158, 216.
48. Drexler (1997), 46; for extensive growth in fixed-shop retailing and fundamental structural change in small trading occurred from the mid-nineteenth century see Scola (1992), chapter IX and Spiekermann (1999), 14.

BIBLIOGRAPHY

Behringer, Wolfgang, 'Reisen als Aspekt einer Kommunikationsgeschichte der frühen Neuzeit', in: Michael Maurer (ed.), *Neue Impulse der Reiseforschung* (Berlin, 1999), 65–95.

Bopp, Matthias, *Die Versorgungslage der Bevölkerung in der Agglomeration Zürich unter besonderer Berücksichtigung des Lebensmittel-Detailhandels* (University of Zurich Ph.D., 1991).

Clark, Peter, *The English Alehouse. A Social History 1200–1830* (London, 1983).

Drexler, Toni, *Kellnerin, a Mass! Das Wirtshaus – die weltliche Mitte des Dorfes* (Fürstenfeldbruck, 1997).

Ebel, Johann G., *Anleitung, auf die nützlichste und genussvollste Art in der Schweitz zu reisen*, 2 parts (Zurich, 1793).

Gercken, Philipp W., *Reisen durch Schwaben, Baiern und die angränzende Schweiz*, 4 vols (Stendal, 1783–1788).

Hersche, Peter, 'Die Lustreise der kleinen Leute – zur geselligen Funktion der barocken Wallfahrt', in: Wolfgang Adam (ed.), *Geselligkeit und Gesellschaft im Barockzeitalter* (Wiesbaden, 1997), 321–332.

Hirschfelder, Gunther, 'Bemerkungen zu Stand und Aufgaben volkskundlich-historischer Alkoholforschung der Neuzeit'. *Rheinisch-westfälische Zeitschrift für Volkskunde* 39 (1994), 87–127.

Holzer, Niklaus A., *Beschreibung des Amtes Laupen 1779*, ed. by Hans A. Michel (Bern, 1984).

Hürlimann, Katja, 'Öffentlicher Konsum in Wirtshäusern. Soziale Funktion des Konsums

in den Zürcher Landvogteien Greifensee und Kyburg im 15./16. Jahrhundert', in: Tanner, Jakob et al. (eds), *Geschichte der Konsumgesellschaft: Märkte, Kultur und Identität (15.–20. Jahrhundert)* (Zurich, 1998), 147–163.
Kerntke, Wilfried, *Taverne und Markt. Ein Beitrag zur Stadtgeschichtsforschung* (Frankfurt a.M., 1987)
Kümin, Beat, 'Useful to have, but difficult to govern. Inns and taverns in early modern Bern and Vaud'. *Journal of Early Modern History* 3 (1999), 153–175.
Kümin, Beat & Radeff, Anne, 'Markt-Wirtschaft. Handelsinfrastruktur und Gastgewerbe im alten Bern'. *Schweizerische Zeitschrift für Geschichte* 50 (2000), 1–19.
Meyr, Karl Georg, *Sammlung der Kurpfalz-Baierischen allgemeinen und besonderen Landes-Verordnungen*, 6 vols (Munich, 1784–1799).
Meyer, Peter, *Illustrierte Berner Enzyklopädie, Vol. 2: Geschichte* (Bern, 1981).
Potthoff, Ossip Demetrius & Kossenhaschen, Georg, *Kulturgeschichte der deutschen Gaststätte, umfassend Deutschland, Österreich, die Schweiz und Deutschböhmen* (Hildesheim, 1996, 1st edn Berlin, 1933).
Radeff, Anne, *Du café dans le chaudron. Economie globale d'Ancien Régime. Suisse occidentale, Franche-Comté et Savoie* (Lausanne, 1996).
Rageth-Fritz, Margrith, *Der Goldene Falken. Der berühmteste Gasthof im alten Bern* (Bern, 1987).
Rennefahrt, Hermann (ed.), *Sammlung Schweizerischer Rechtsquellen. Abt. II: Die Rechtsquellen des Kantons Bern. 1. Teil: Stadtrechte*, vol. VIII/1: Wirtschaftsrecht (Aarau, 1966).
Sandgruber, Roman, *Die Anfänge der Konsumgesellschaft: Konsumgüterverbrauch, Lebensstandart und Alltagskultur in Österreich im 18. und 19. Jahrhundert* (München, 1982).
Schärer, Martin R., 'Food history in Switzerland. A survey of the literature', in: H.-J. Teuteberg (ed.), *European Food History. A Research Review* (Leicester, 1992), 168–198.
Schremmer, Eckhart, *Die Wirtschaft Bayerns. Vom hohen Mittelalter bis zum Beginn der Industrialisierung* (München, 1970).
Schuck, Gerhard, 'Bayern', in: Karl Härter and Michael Stolleis (eds), *Repertorium der Policeyordnungen der frühen Neuzeit, Vol. 3: Wittelsbachische Territorien*, Vol. 3/1 (Frankfurt a.M., 1999), 595–613.
Scola, Roger, *Feeding the Victorian city: The food supply of Manchester, 1770–1870*, ed. by W. A. Armstrong & Pauline Scola (Manchester, 1992).
Spiekermann, Uwe, *Basis der Konsumgesellschaft. Entstehung und Entwicklung des modernen Kleinhandels in Deutschland 1850–1914* (Munich, 1999).
Stürler, Moritz von, 'Die konzessionierten Wirtschaften im Kanton Bern'. *Berner Zeitschrift für Geschichte und Heimatkunde* 37 (1975), 19–27.
Tanner, Jakob et al. (eds), *Geschichte der Konsumgesellschaft: Märkte, Kultur und Identität, 15.–20. Jahrhundert* (Zurich, 1998).
Teuteberg, Hans-Jürgen, 'Homo edens. Reflexionen zu einer neuen Kulturgeschichte des Essens'. *Historische Zeitschrift* 265 (1997), 1–28.
Thurner, Adolf, 'Der "Alte Wirt" in Obermenzing'. *Amperland* 22 (1986), 268–271.
Zeiss, Hans, 'Zur Geschichte der Tafernen in Altbayern', in: K. v. Manz et al. (eds), *Heimatarbeit und Heimatforschung*. Festgabe für Christian Frank zum 60. Geburtstag (München, 1927), 203–210.

Part 2
The urban-rural interface

JOHN BURNETT

7 The English village labourer revisited

Self-provisioning and the commercialization of diet, 1790–1914

In 1911 J. L. and Barbara Hammond published what was to become a founding text in the emerging study of economic and social history. *The Village Labourer* was an examination of the effects of the enclosure movement in agriculture which transformed many of the open fields and commons in midland and southern England into consolidated holdings which permitted more efficient land use and the adoption of new techniques of intensive farming. The Hammonds argued that in this process the labourer, who had formerly enjoyed customary rights to small plots of land and to the use of commons and wastes for grazing animals, almost always lost out to large landowners who had the knowledge and resources to process the complexities of enclosure settlements to their own advantage. Former peasant proprietors who had been largely self-supporting, unable to substantiate their claims, were reduced to hired labourers dependent on wages which in the later eighteenth century increasingly lagged behind the prices of the necessities which they now had to buy. The authors concluded their account of the emiseration of the English rural labourer to a level of bare subsistence, often dependent on poor relief, with a graphic description of the widespread riots of 1830–1831, which they titled The Last Labourers' Revolt.[1]

Sympathetic observers of rural poverty had little doubt that the root cause was enclosures and the engrossing of land into large, capitalist estates run for profit. "An amazing number of people have been reduced from a comfortable state of independence to the precarious position of mere hirelings", complained the Revd. David Davies in 1795. In the past, wages had been based on the understanding that the labourer had other means of support and were, to an extent, supplementary: now the loss of the commons deprived his family of milk, cheese and butter, of keeping a pig and poultry and of gathering free fuel. The remedy, Davies and other well-wishers believed, was to restore the labourer to a little land. "Give him the ability to keep a cow, and then all will have milk. Give to all the ability to brew small beer, and then few will frequent ale-houses".[2]

A romantic myth of a 'golden age' of self-sufficiency before enclosures lingered throughout the nineteenth century. Memory doubtless idealized the past, but it was true that both the economic structure and the social fabric of the English village were largely transformed by the commercialization of agriculture into one in which the profit motive and the cash nexus predominated. A year after the Hammonds' book appeared, George Sturt published his *Change in the Village* (1912), an account of the altered circumstances in a small Surrey village where the common was not enclosed until 1861. Older inhabitants remembered cutting turf for fuel, growing corn, baking bread and brewing beer. But afterwards, "the once self-supporting cottager turned into a spender of money at the baker's, the coal-merchant's, the provision-dealer's, and, of course, needing to spend money he needed first to get it [...] Every week money has to be found, and not only, as of old, for rent and boots and for some bread and flour, but also for butter or margarine, sugar, tea, bacon or foreign meat if possible, lard, jam, and – in the winter at least – coal".[3]

Recent scholars of agrarian history have refined the Hammonds' general condemnation of enclosures, which had different effects for different groups – small farmers, cottagers and squatters – and were not universally disastrous. Some benefited by exchanging scattered strips of land for compact holdings, some found extra paid work on large, intensively-cultivated estates, at least during the boom years of the French Wars (1793–1815). The loss of the commons was most acutely felt after 1815 when prosperity turned into depression, and an increasing rural population struggled to find regular employment, even at starvation wages. But to suppose that before the nineteenth century most rural workers were small peasant farmers possessing land and animals is a myth: it is estimated that as far back as 1690 there were already 1.74 landless labourers to every occupier of land and that in 1831, by which time the great wave of enclosure (c.1760–1830) was over, the proportion had increased only to 2.5 landless labourers to each occupier.[4]

What is not disputed is that during the nineteenth century England, uniquely in Europe, became a country of large farms whose ownership was increasingly concentrated into fewer hands. English agriculture was characterized by a tripartite, hierarchical division between a small number of very large landowners, many of aristocratic lineage, a bigger group of tenant-farmers paying rent and working the land by hired labour, and a large class of mainly landless labourers dependent on wage-earning. Social reformers were concerned at the growing 'monopoly' of land ownership, apparently confirmed by an official 'New Domesday Survey' in 1873 showing that two-thirds of the 33 million acres[5] of land in England and Wales were in estates of 500 acres or more, and that four-sevenths of the whole area were owned by 4,000 persons.[6] No significant change had occurred by 1913 when another enquiry estimated that only one per cent (285,000 acres) of English and Welsh land was in small holdings of between one and five acres.[7]

This chapter first examines the extent to which labourers were able to provision themselves in the nineteenth century, mainly through the development of the allotment system, and secondly reviews the ways in which their diets became increasingly commercialized and dependent on purchased foods.

The allotment movement

Throughout the nineteenth century the agricultural labourer was the lowest-paid of all regularly-employed workers. In 1824 in some southern counties wages were as low as 4s. 6d.[8] a week for a married man, in others 8s. or 9s., and in some northern areas, where competition from industries and mining pushed up the level, as much as 12s. or even 15s.[9] This dichotomy between the arable south and east and the more pastoral north and west persisted throughout the century, although over time the gap closed somewhat: in 1850–1851 the average agricultural wage was 9s. 6d. a week, 8s. 5d. in the south and 11s. 6d. in the north.[10] Improvement dated mainly from the 1880s when food prices fell and wages rose as labourers made themselves a scarcer commodity by migration to towns and colonies: by 1902 average family earnings had grown to an estimated 18s. 6d. a week.[11] Improvement there had been, but from such a low base that many families still existed in poverty, now scientifically defined in nutritional terms by Seebohm Rowntree:[12] in Corsley, Wiltshire, in 1909, 29 of the 70 labourers' families were below this level of "merely physical efficiency",[13] and as an Oxfordshire wife, struggling to maintain a family of five children under the age of six on 12s. a week, commented bitterly to Rowntree: "You can't call it living; it's a dragging of yourself along".[14]

The apparent impossibility of raising wages to a satisfactory living standard lay behind the economic argument for providing labourers with small plots of land on which they could grow some of their food requirements. But the allotment movement was always viewed in more than purely material terms. Some protagonists regarded it as a just return for the loss of land by enclosures: many believed that a stake in the land would restore the labourer to independence, free from the demoralising shackles of the poor law, while others again argued that a little land and a decent cottage would provide the foundations for a contented, dutiful peasantry. It is noteworthy that allotments were most actively promoted as panaceas for social discontent at times of crisis in the labourer's condition. Thus, an initiative by several large landowners dates from the disastrous harvest of 1795 when prices soared and food riots occurred in many parts of the country. Some commentators like Sir Frederic Eden believed that poverty was due to the labourer's extravagant taste for white bread and tea, and recommended a return to cheaper grains such as barley, rye and oats,[15] but there were good reasons why the earlier dietary pattern, which had included meat, dairy products and vegetables as well as cereals, was not now possible. A more practicable remedy was to encourage the labourer to grow some of his own food, one of the policies advocated by The Society for Bettering the Condition and Improving the Comforts of the Poor, established in 1796 under royal patronage. The provision of allotments by philanthropic landowners grew slowly through the years of the French Wars, but although there were many reports of their good effects – that labourers became more "steady and trusty", sober and contented and rarely applied for poor relief – farmers were often unwilling to set land aside for such purposes. Possession of land, it was said, would make labourers too independent, "too saucy": they would breed more children – a particular Malthusian fear at this time: they would be too exhausted from tending their own allotments to give their employer a fair day's work and would not be available for extra work when the farmer most needed them, for example at harvest-time. Farmers often looked more favourably on potato-grounds which they let on the fallow fields for the season than on more permanent allotments: potatoes, carefully cultivated by the spade and manured, cleaned and improved the land for next year's cereal crop.

Allotments were more actively promoted during the agricultural depression after 1815, and were now officially endorsed by legislation. The Select Vestries Act, 1819, empowered, though did not require, parishes to buy or lease up to 20 acres to let to poor, industrious labourers in quarter-acre plots: further Acts in 1831 and 1832 – significantly passed after The Last Labourers' Revolt – raised the amount to 50 acres, but it appears that the influence of farmers in parish affairs prevented any general adoption of these measures. No national statistics of the extent of allotments at this time exist, but it is clear that provision was greater in the low-wage counties of the south and east than in the north, apparently corresponding with the degree of poverty. Wiltshire, one of the poorest areas of the country, probably had the best provision of all, a survey of 26 parishes there in 1833–1834 showing that 1,450 cottagers had allotments out of a labouring population of 2,200.[16] This association of the allotment movement with the problem of poverty and its treatment was very evident in the enquiries carried out by the Poor Law Commissioners in 1833. They believed that the effects of allotments were highly beneficial, and that nothing could do more to restore "the good feeling which formerly prevailed between the different classes of society".[17] The best size, they thought, was a quarter of an acre, which could be cultivated by a man and his family in their leisure time: to grant more would make him "not a valuable servant" – even so, men were known to have worked their plots by candlelight after twelve hours' labour.[18] It was also reported that labourers were sometimes suspicious of allotments let by the parish, and at Warminster, where 20 acres were offered on condition that the labourers forfeited their right to poor relief, there were no takers:[19] poor relief was one of the few rights left to them, and a better safety-net than a small piece of land which needed their extra time and effort.

Access to land was still very restricted in the 1830s when the Commissioners attempted to survey the provision. Allotments were reported in just under half of midland and southern counties, but there was no indication of their numbers or size: cottage gardens were possessed by all or nearly all labourers in 57 per cent of southern parishes and far fewer in the north,[20] a surprise to those who had assumed that every cottager at least had a small patch of land.

Between 1795 and 1835 no fewer than 184 pamphlets on the subject of land for the labourer were published, and allotments continued to be proposed as a favourite remedy for rural poverty. A Select Committee in 1843 gave further support to them, but stopped short of recommending any compulsory provisions since the rights of private property were regarded as paramount. The same difficulty was encountered in the General Inclosure Act, 1845, which sought to ensure that allotments would be provided in all future enclosures, yet of 614,800 acres enclosed between then and 1869 only 2,223 acres (one three-hundredth part) were so set aside.[21] The allotment movement did not make greater progress because although it had official encouragement it also had some strong opposition from people directly involved. 'Official' opinion, represented in Select Committees and Royal Commissions, generally took the view that a quarter of an acre planted with potatoes and other vegetables could make all the difference to the labourer's standard of living between bare subsistence and tolerable comfort. Precisely how much an allotment contributed depended on its size, fertility, the skill of the labourer and the amount of help from his wife and children. A Cambridgeshire budget of the 1830s indicated a value of about £5 a year or 1s. 9d. a week, adding around a fifth to the wage,[22] but a much lower estimate from Bedfordshire in 1867 placed the annual value at between £1 and £1. 10s. 0d. when rent and other expenses were deducted:[23] in 1913 Rowntree and Kendall's survey of labourers' budgets suggested that the quantity of home-produced food ranged between 10 per cent and 25 per cent. Whatever the exact proportion, self-provisioning was clearly a major influence on the labourer's diet. "Potatoes are the staple food, which are relied on to eke out bread [...] but families are seldom supplied with potatoes the whole year round. As for the more nutritious garden stuff such as peas or beans, it is regarded as more or less of a luxury".[24] The prejudice against potatoes which William Cobbett and others had voiced earlier had long since disappeared, partly because food that the labourer grew himself carried old associations of pride and independence. "The best thing in them days was the 'lotments, said an old labourer remembering the 1840s, while another replied to a question:

> You ask 'ow the people did get on. Well, they got into debt, and then again, they lived on 'taters' and kept pigs, but butcher's meat we never 'eard of, never saw it except in the shops.[25]

By 1850 there may have been around 100,000 allotments, mainly in the midlands, East Anglia and parts of Southern England:[26] this would suggest that perhaps one in 10 of the 965,000 agricultural workers possessed one, though allowing for unmarried sons of working age living at home, nearer one in four families. But, as Jeremy Burchardt has shown, opinions were by no means universally favourable. Although many landowners were sympathetic, some, imbued with the principles of the Classical Economists, opposed any excessive subdivision of the land, fearing that allotment-holders would be tempted to early marriage and become an under-capitalized poverty-stricken 'cottier' class as in Ireland. Farmers were generally less favourable, often opposed to the inconvenience of letting small plots from their fields even at excessive rents. More fundamentally, many feared that 'their' labourers would exhaust their strength on their allotments, would become too independent and weaken the farmers' bargaining position over wages and working hours. Nor can it be assumed that labourers themselves were always enthusiastic about allotments, especially the conditions on which some were let. Resentment arose over excessive rents, inadequate size (one-eighth of an acre was common on some estates), remoteness from the cottage, control over the crops to be grown and the forbidding of Sunday working – the labourer's only free day.[27]

Nevertheless, the provision of allotments continued to grow throughout the second half of the century. In 1871 it was estimated that there were now 250,000 allotments under a quarter of an acre and a further 160,000 above that size, but the compiler of this Census was unable to distinguish between allotments held by agricultural labourers and those occupied by other classes, including a growing number in urban and suburban areas. Pressure in Parliament by Jesse Collings, the Liberal M. P. for Ipswich, resulted in an Allotments Extension Act in 1882, which aimed to compel the trustees of charitable lands to offer them as allotments. The difficulties and delays involved in the procedure were well illustrated at Tysoe in Warwickshire, where the labourers had a long struggle with the trustees of the ancient Town Lands Charity to allow it to be used for allotments.[28] Another Act in 1887 allowed local authorities to buy or rent land to be laid out as allotments, but by this time interest was turning to a larger issue than small vegetable plots – to the possibility of smallholdings of perhaps 5–20 acres which would provide self-sufficiency for a family as in many European countries. This was in part a response to the deepening agricultural depression which began in the 1870s when arable farmers had to compete with much cheaper wheat imports from North America: land prices and rents fell dramatically, while several million acres went out of cultivation. Some observers began to propose smallholdings as a means of agricultural revival, providing family employment in specialized branches such as dairying, market gardening, poultry-rearing and fruit-growing. "The most obvious solution to the agricultural difficulty would be to win back the labourer to the land", declared James Thorold Rogers in 1884: it would, moreover, halt "the rural exodus [by which] the poverty of the towns has been swollen".[29] A new element in the debate was the increasing political influence of labourers through the formation of agricultural trades unions and the extension of the parliamentary franchise to the rural working class in 1884. The labourer's voice, so long silent, was now being heard, and, as a voter, he was being courted by the main political parties and by newly-formed radical groups which aimed at a transformation of the monopoly of land-ownership. Villages were visited by brightly-coloured vans from neighbouring towns from which speakers proclaimed their various remedies for the problems of the day: they included the Home Rule (for Ireland) Van, the Beaconsfield (Conservative) Van, the Sunrise Radical Van, the Yellow Van of the Land Nationalization Society and the Red Van of the Land Restoration League.[30] The outcome of parliamentary pressure for smallholdings led by Joseph Chamberlain, Jesse Collings and Joseph Arch, the first farm-worker to be elected in 1885, was a tentative Act in 1892 which allowed the new County Councils to establish smallholdings for sale provided the applicant made a 20 per cent deposit.

Only 652 acres were acquired during the following 10 years, and in 1906 a Departmental Committee reported that although there had been some enthusiastic support in a few areas, particularly in Norfolk and Lincolnshire, many labourers had not shown interest because they were unable to raise the necessary capital and often lacked appropriate skills for this type of farming: others again feared victimization by their employer if their application was unsuccessful. The Liberal government's Small Holdings and Allotments Act, 1907, was intended to offer a final solution to what had become the contentious issue of the labourer's access to land. It now became the responsibility of County Councils to investigate the demand for smallholdings and to draft schemes to meet it: they could acquire land, if necessary by compulsory purchase, and then either sell smallholdings by instalments of up to 50 years or let them to those with insufficient capital. By 1914 Councils had acquired 200,000 acres, and there was something of a rush of applicants for 14,000 smallholdings, but only two per cent of farmworkers applied and many were refused, mainly for lack of capital.[31] Of all successful applicants only one-third were agricultural labourers:[32] the majority went to sons of small farmers, village craftsmen, publicans, carriers and the like.

If smallholdings had not made any dramatic change in English landownership by 1914, labourers were generally much better supplied with allotments than a century earlier. Nor did the vegetables which they grew there and in their cottage gardens exhaust the possibilities of self-provisioning. Many managed to rear a pig a year, and pig-meat, in the form of bacon, ham and offal, was the most usual kind of meat which families tasted. "The pig is to the farm labourer a kind of savings bank" wrote Francis Heath,[33] but he needed some capital to open his account: a piglet cost between 10s. and 25s., and thereafter there was barley-meal to be bought in addition to the kitchen pig-swill and the 'pig-nuts' (acorns) which the children collected. It was common for half the pig to be mortgaged in advance to the meal-man, or, as at Quarry Hill, near Oxford, for two pigs to be reared, one for the family and one to pay debts.[34] Nevertheless, pig-killing time in the winter was an event for celebration, with a Sunday 'pig-feast' when relatives, friends and children away in service assembled for a rare indulgence.[35]

In many arable areas gleaning ('leazing') was allowed in the autumn, when women and children picked the remaining ears of corn left by the horse-rake: this could provide flour for several months – not, usually, for bread, which was bought, but for puddings, pies and dumplings. Some men poached rabbits and game-birds, partly for food and partly for the thrill of the chase, though they ran a terrible risk of fine, imprisonment and, in the earlier century, even transportation: as late as 1880 22 per cent of the crimes tried at Henley Petty Sessions was for poaching.[36] There were also often some legitimate 'perquisites' which could affect the labourer's dietary standards. Some farmers gave free meals as a spur to effort at busy times: in dairying regions skim milk was sometimes given or sold at a nominal price, while free beer for men at work was widespread. The quantity and strength varied considerably. In some areas 'small beer' (a second mashing of the same malt) was given throughout the year, in others, strong beer during certain tasks – in Cambridgeshire at haymaking and harvest a pint at each of the five meal-breaks, in Bedfordshire four or five pints a day during sheep-shearing.[37] In the western counties cider, locally made on hundreds of farms, took the place of beer, and was 'given' at up to 6 pints a day, though some friends of the poor objected that this was simply a concealed 'truck' payment: in Devon, Canon Girdlestone was indignant that wages in the 1870s were only 8s. or 9s. a week, plus 4 pints of cider a day, "very washy and sour [...] valued at 2s. a week, but much over-valued".[38]

The growth of a commercialized diet

Throughout the nineteenth century a common view persisted that however low his wages, the rural labourer had the advantages of access to land and a measure of self-sufficiency which provided him with a wholesome diet, more 'natural' than that of the town worker who depended entirely on bought food, often of inferior quality. Although this view was largely the product of literary invention, labourers themselves tended to recall a time not dominated by the market economy. An old Hampshire woman in 1902 nostalgically remembered her childhood thus:

> Nothing's good enough now unless you buys it in a public house or a shop. It wasn't so when I was a girl. We did everything for ourselves, and it were better, I tell 'e. We kep' a pig then, so did everyone, and the pork and brawn it were good, not like what we buy now [...] And we didn't drink no tea then [...] We had beer for breakfast then, and it did us good [...] It were better than all these nasty cocoa stuffs we drinks now [...] And we had a brick oven then, and could put a pie in and a loaf and whatever we wanted, and it were proper vittals.[39]

If memory did not play her false, her experience was quite untypical. By the mid-nineteenth century tea was the all but universal drink of labourers' families and beer, unless given at work, was a man's evening luxury: by no means all could afford to keep a pig, and many cottages lacked an oven or any cooking facilities except an open fire. Already by the late eighteenth century most of the food of labourers, especially in the south of England, was shop-bought, as the Revd. David Davies discovered in his parish of Barkham, Berkshire, in 1787: "I found them in general but indifferently fed [...] and most families in debt to little shopkeepers." The women still baked their bread, but bought the flour, salt and yeast which, in some budgets, took three-quarters of the whole earnings: typically there was in addition for a week one pound of bacon, 8d. ("boiled two or three times with greens"); one ounce[40] of tea, 2d.; three-quarters of a pound of sugar, 6d. and half a pound of butter or lard, 4d. With fewer children (this family had five), or children of working age, there was more tea and bacon and a little cheese, but the primary dependence on bread was the same in all cases. Only two of the Barkham families occasionally brewed a small quantity of beer, usually for a christening celebration, and milk was rarely mentioned because "Suckling is here so profitable (to furnish veal for London) that the poor can seldom either buy or beg milk".[41]

Respondents to Davies' enquiries in the north of England showed a rather different picture. In Westmorland there was milk in all the budgets: the bread was made of barley and rye, the porridge or 'hasty-pudding' of oatmeal and there were plenty of potatoes: in Co. Durham some families spent as much as 1s. 6d. a week on milk.[42] Another advantage in the north was a greater availability of fuel – furze, heath and brushwood – which allowed more cooking of soups and stews, more home baking and less dependence on bought foods. Nevertheless, in all the budgets collected at the time by David Davies and Sir Frederic Eden,[43] which covered the whole country, there are some interesting similarities. Carole Shammas concludes that in both north and south 11.2 per cent of food expenditure went on tea and sugar and only 2.5 per cent on beer, but on dairy products northern labourers spent 16 per cent compared with five per cent in the south:[44] the northern diet provided substantially more energy – 2,352 kcals per person per day (2,823 per adult) compared to 1,734 kcals (2,109 per adult) in the south, well below the requirement for a man engaged in heavy work.[45]

Well before the nineteenth century the age of self-sufficiency, if it ever existed, was long past. Even in the seventeenth century it had been commented that "Now in every country village where is not above ten houses there is a shopkeeper",[46] but their numbers grew rapidly in the next century under the impact of rising population and demand. Groceries, especially tea, coffee and sugar, were the fastest-growing sector of retail trade, so much so that in 1783 there were 52,292 licensed tea dealers in England and Wales, at least one to 180 people.[47] The petty village shop supplied a wide range of the day-to-day wants of local inhabitants. At the lowest level the 'shop' was merely the front room of a cottage, often kept by a wife or widow. In a village of greater size, like East Hoathley, Sussex, Thomas Turner acted as grocer, draper, haberdasher, hatter, druggist and ironmonger as well as keeping a small school,[48] while at Kirkby Stephen in Westmorland, more of a small town than a village (population 1,141 in 1801), Abraham Dent's was only one of ten shops, seven of which sold groceries: his stock included tea (4s. 8d. – 19s. 0d. per pound), sugar (4d. – 1s. 0d. per pound), coffee (4s. – 6s. per pound), cocoa (1s. per pound), flour, hops, rice, dried fruits and spices, but no butter, eggs or cheese, which were available at the weekly market. His customers ranged from clergy and doctors down to labourers, to whom he would sell one ounce of tea and one-quarter of a pound of sugar and give credit.[49]

In the nineteenth century the village shop provided a low-cost distribution system for a population which included many people living at or little above subsistence level. The shopkeeper had minimal fixed costs, employed no staff, and often had other sidelines to his business. He bought from wholesalers in neighbouring towns or from middlemen who handled the imported grocery products and were already well established in English commerce.[50] His main risks were not so much in his stock, which had a relatively long shelf-life, as in bad debts, which in Dent's case ran from one shilling to seven guineas. Perishable foods like meat, fresh vegetables, fruit, butter, cheese and eggs were sold in weekly or twice-weekly markets, supplied from local sources. With at least 700 markets officially recorded, most villagers lived within walking distance or could pay a small sum for a lift on a carrier's cart. Shops and markets were, therefore, complementary rather than competitive, though market towns also had a greater range of more specialised shops which might be patronized on visits to town. As Mary Mitford described it in 1824, the village shop was "multifarious as a bazaar, a repository for bread, shoes, tea, cheese, tape, ribands and bacon, for everything, in short, except the one particular thing which you happen to want…"[51] Its necessarily limited stock was supplemented by the visits of itinerant traders such as pedlars and packmen, who carried low-bulk items on their backs or in pushed carts - not usually food, but with the exception of tea, which was widely carried by the curiously-named "Scotch drapers". A somewhat distinct category was the hawker, who paid regular, often weekly, visits to villages, and generally kept a stall in the local market: he brought vegetables, fruit and fish, usually cured herrings or bloaters, although, at Heyshott, Sussex, a man stocked up with fresh fish in Chichester and hawked it round in a little cart drawn by two dogs. The 1851 Census showed 25,747 pedlars, hawkers and hucksters, not including occasional traders who often took up the work in periods of unemployment.[52]

Labourers' budgets in the mid-nineteenth century showed little change from those described by Davies and Eden. In 1843 a Sussex labourer with five children aged under 12 years spent the whole of his wage of 9s. a week on bread alone: three children earned 4s. between them and his wife averaged 9d. a week by field-work over the year: after rent (1s. 2d), they bought potatoes (1s.), butter and cheese (7 ½d.) and tea and sugar (5 ½d.).[53] There was little incentive here for the village shop to stock anything beyond basic necessities. Ten years later *The Shopkeeper's Guide* (1853) listed the recommended items for such a shop as tea, coffee, cocoa, chicory, flour, barley, semolina, pepper, mustard, spices and sauces:[54] sugar was omitted, presumably due to an oversight, since this was a basic item in any grocery.

A transformation of at least some village shops may be dated from the 1870s. By then, many had added the function of a sub-Post Office following the introduction of the national postal service in 1840 and the growth of literacy in the countryside: the shop even served for the weekly visits of the Relieving Officer to distribute outdoor relief to the paupers. It has been described as "the most central focal-point of common intercourse and activities".[55] The important change affecting the village shop was the rise in real incomes which agricultural labourers shared with other workers in the later century,[56] and the increased consumption of many foods which this made possible: between 1860 and 1900 national consumption of tea rose from 2.7 to 6.1 pounds per year, sugar from 34.1 to 87.1 pounds. As Maude Davies noted in Corsley in 1909: "There has been a great increase in the quantity of groceries consumed by cottagers within the last forty years, since the rise of wages".[57] An official enquiry in 1902 estimated the average earnings of labourers' families as 18s. 6d. a week, with 13s. 6 ½d. spent on food: more now went on meat (fresh, pork and bacon) than on flour and bread, instead of one ounce of tea nearly half a pound was consumed, four and a quarter pounds of sugar, one pound of butter and one and a quarter pounds of cheese.[58] Recalling village life in the 1850s, a labourer wrote in 1905: "Within my recollection there was scarcely a trade cart came into the village to call on labourers [...] Go into the same village today, and you may have to be on the look-out not to be run down by the vehicles of tradesmen and others. Yet the inhabitants are, I believe, considerably less".[59]

In the later nineteenth century the village began to experience something of 'the retailing revolution' which was transforming High Street shopping in the towns. This was especially evident in groceries where multiple chains like Lipton's, the Home and Colonial and the Maypole Dairy Co. were responding to a strong working-class demand on the principles of bulk purchase, high turnover and low prices.[60] The rapidly-expanding co-operative retail societies were performing a similar function for their 1,707,000 members in 1900, with the added inducement of a dividend on purchases. Multiple trading depended on a mass demand and did not encompass the village, but few market towns by 1914 were without branches of the stores, which were already household names. Another important commercial trend was the emergence of large-scale firms specializing in the manufacture or distribution of particular food or drink products such as tea, cocoa, sugar, jam, milk (both fresh and condensed) and margarine. These were branded goods, packaged at source, widely advertised and sold at fixed prices, releasing the shopkeeper from blending teas, curing bacon, grinding coffee and breaking up sugar loaves: if food was becoming more commodified and depersonalized, the customer could generally be assured of reliable quality and competitive prices.

Labourers' budgets collected before 1914 show that the range of bought foods had expanded greatly since the restricted diets of mid-century. In Corsley one of the poorest labourers earning only 15s. a week to support a wife and five young children bought butter, cheese, Quaker Oats (rolled oats for quick preparation of porridge), suet (for pies and puddings), currants, oranges, one shilling's worth of beef besides bacon and a little fish.[61] Other budgets of the period include coffee as well as tea, Fry's Cocoa, bloaters, rabbits (6d.), egg powder, "Camp Coffee" (an early 'instant'), "Extract of Beef", Swiss condensed milk, margarine, jam, Fray Bentos corned beef, "Golden Syrup", potted meat, sausages, ginger ale and fruit jellies; occasional 'treats' were canned fish (salmon and sardines) and tropical fruit. Although many of these diets still fell short of the "physical efficiency" test, especially in large families of children, they were much more varied and palatable than formerly. Apart from home-grown vegetables and whatever fruit might be gathered, they were essentially purchased diets, which were becoming more standardized, less distinct regionally and also less distinct from urban food patterns.

Although it now faced competition from multiple and co-operative stores in nearby market towns – now more accessible by branch railways and organized carrier services – the village shop survived by offering personal services, credit (which multiples did not) and willingness to sell in tiny quantities. But it was now economically worthwhile for town grocers, bakers, butchers, fishmongers and dairymen to provide regular delivery services to villages. Town co-operative societies enrolled rural members, supplying them by travelling shops[62] or despatching orders by carriers.[63] Much had changed in the labourer's circumstances since the privations of a century earlier, and even George Sturt, who idealized the old way of life (from a middle-class perspective) could see that there had been improvements in living standards and social relationships:

> It is quite the common thing in the villages for a man and wife to lock up their cottage on a Saturday evening, and go off with the children to do the week's shopping together. On a nice night the town becomes thronged with them, and so do the shops [...] The majority of the people make a little weekly festivity of this Saturday night's outing; they meet their friends in the street, have a chat, wind up with a visit to the public-house, and so homewards at any time between seven and ten o'clock, trooping up the hill happily enough as a rule.[64]

NOTES

1. Hammond & Hammond (1948). The revolt is now usually described as the "Swing Riots" after a supposed leader, "Captain Swing". The more modern account is by Hobsbawm & Rudé (1973).
2. Davies, D. (1977), 40, 57.
3. Sturt (1984), 9, 89–90. My Introduction to this edition gives a brief account of Sturt's life and his best-known book, *The Wheelwright's Shop* (1923).
4. Chambers & Mingay (1966), 103. This contains a good review of theories on the effects of enclosures in ch. 4, 77–106. More recent judgements are in Armstrong (1988), 33–43.
5. One acre = 4,047 m2 (or 0.4 hectares).
6. Brodrick (1968). Table p. 158 et seq., 164–165.
7. *The Land* (1913), 157, 171.
8. The pound sterling (£1) is divided into 20 shillings (s.) and each shilling is divided into 12 pence (d.). A guinea is 21 shillings (£1. 1s.).
9. *Report from the Select Committee on the Rate of Agricultural Wages, and on the Condition and Morals of Labourers in that Employment* (1824).
10. Caird (1852).
11. Board of Trade (1903), 211.
12. Rowntree defined poverty as living below a level of physical efficiency, which, for a man in moderate work, required 3,500 kcals and 125 gs of protein per day.
13. Davies, M. F. (1909), 149.
14. Rowntree & Kendall (1917), 31–32, 50.
15. Eden (1966), vol. 1, 495.
16. Barnett (1967), 169–170.
17. *Report on the Administration and Practical Operation of the Poor Laws* (1834), 182–183.
18. Ibid., 186–187.

19. Ibid., 190.
20. Barnett (1967), 171–172.
21. Burchardt (1997), 166.
22. Barnett (1967), 169.
23. Agar (1981), 23.
24. Rowntree & Kendall (1917), 45, 50, 62, 331.
25. Unwin (1905), 11, 16.
26. Burchardt (1997), 166.
27. Horn (1987), 8, for details of restrictive covenants on allotments on Lord Tollemache's estate at Helmingham, Sussex.
28. Ashby (1974), 117–133.
29. Rogers (1884), 561, 564.
30. The Land Restoration League received financial support from the recently-formed dockers' trade union in the hope that it would reduce migration to the towns.
31. Armstrong (1988), 148.
32. *The Land* (1913), 193.
33. Heath (1874), 113.
34. Samuel (1975), 190.
35. Thompson (1973), 24–27. The account relates to a village in N. Oxfordshire in the 1880s.
36. Horn (1987), 222.
37. Horn (1980), 29–30: Agar (1981), 11.
38. Heath (1874), 99–100. Payment of wages in kind was, in theory, made illegal by the Truck Act, 1887, but was often unenforced.
39. Fussell (1949), 128.
40. One ounce is equal to 28.35 g and there are 16 ounces (oz.) in one pound (lb.).
41. Davies D. (1977), 6–19.
42. Ibid., 156, 184, 187.
43. Eden (1797, 1966).
44. Shammas (1984), Table 1, 257.
45. Shammas (1990), 134–135.
46. Anon (1681), 397, quoted in Mui & Mui (1989), 149.
47. Mui & Mui (1989), 161–164.
48. Jennings (1979), xxx. Turner was born in 1729; his diary begins in 1754.
49. Willan (1970), 1–25, 153–155.
50. Alexander (1970), 6, citing Westerfield (1915), 124–125.
51. Mitford (1893), 10–11.
52. Alexander (1970), 61–64, 77–78.
53. *Report on the Employment of Women and Children in Agriculture* (1843), 233.
54. Anon (1853), cited Davis (1966), 267–268.
55. Fletcher (1977), 40–41. Richard Cobbold was Rector of Wortham, Suffolk, from 1825–1877: he spent £70–80 a year in the village shop.
56. The index of real incomes per head at 1900 prices (1900 = 100) rose from 57.7 in 1870 to 63.2 in 1880, 89.4 in 1890 and 100 in 1900. Jeffereys (1950), 9, footnote.
57. Davies, M. F. (1909), 194–195.
58. Board of Trade (1903), 211. Statistics relate to the "average" family of husband, wife and 4 children.
59. Unwin (1905), 35.
60. See generally, Mathias (1967).
61. Davies, M. F. (1909). Budget no. 44, 196–197.
62. Birchall (1994), 118–119.
63. Griffiths (1979), 60.
64. Sturt (1984), 30–31.

BIBLIOGRAPHY

Agar, Nigel E., *The Bedfordshire Farm Worker in the Nineteenth Century*, Bedfordshire Historical Record Society, vol. 60 (London, 1981).
Alexander, David, *Retailing in England during the Industrial Revolution* (London ,1970).
Anon., *Trade of England Revised* (1681), quoted in Mui & Mui (see under).
Anon., *The Shopkeeper's Guide* (1853), quoted in Davis, Dorothy (see under).
Armstrong, Alan, *Farmworkers. A Social and Economic History 1770–1980* (London, 1988).
Ashby, M. K., *Joseph Ashby of Tysoe, 1859–1919* (London, 1974).
Barnett, D. C., 'Allotments and the Problem of Rural Poverty, 1780–1840', in: E. L. Jones & G. E. Mingay (eds), *Land, Labour and Population in the Industrial Revolution. Essays presented to J. D. Chambers* (London, 1967).
Birchall, Johnston, *Co-op: the people's business* (Manchester, 1994).
Board of Trade, *Memoranda, Statistical Tables and Charts with reference to British and Foreign Trade, etc.*, Cd.1761 (1903).
Brodrick, George C., *English Land and English Landlords* (New York, 1968, 1st edn 1881).
Burchardt, Jeremy, 'Rural Social Relations, 1830–1850. Opposition to Allotments for Labourers'. *Agricultural History Review* 45 (2) (1997).
Caird, James, *English Agriculture in 1850–1851* (London, 1852).
Chambers, J. D. & Mingay, G. E., *The Agricultural Revolution, 1750–1880* (London, 1966).
Davies, David, *The Case of Labourers in Husbandry, Stated and Considered* (New Jersey, 1977, 1st edn 1795).
Davies, Maude F., *Life in an English Village. An Economic and Historical Survey of the Parish of Corsley in Wiltshire* (London, 1909).
Davis, Dorothy, *A History of Shopping* (London, 1966).
Eden, Sir Frederic Morton, *The State of the Poor*, 3 vols (London, 1966, 1st edn 1797).
Fletcher, Ronald (ed.), *The Biography of a Victorian Village. Richard Cobbold's Account of Wortham, Suffolk, 1860* (London, 1977).
Fussell, G. E., *The English Rural Labourer from Tudor to Victorian Times* (London, 1949).
Griffiths, Winifred, *One Woman's Story* (Rhondda, 1979).
Hammond, J. L. & Hammond, Barbara, *The Village Labourer*, 2 vols (London, 1948, 1st edn 1911).
Heath, Francis George, *The English Peasantry* (London, 1874).
Hobsbawm, E. J., & Rudé, George, *Captain Swing* (London, 1973).
Horn, Pamela, *Labouring Life in the Victorian Countryside* (Gloucester, 1987).
Horn, Pamela, *The Rural World, 1780–1850. Social Change in the English Countryside* (London, 1980).
Jeffereys, James B., *Retail Trading in Britain, 1850–1950* (Cambridge, 1950).
Jennings, G. H. (ed.), *Thomas Turner, The Diary of a Georgian Shopkeeper*, 2nd edn (Oxford, 1979).
The Land, The Report of the Land Enquiry Committee, vol. 1, 3rd edn (London, 1913).
Mathias, P., *Retailing Revolution. A History of Multiple Retailing in the Food Trades based on the Allied Suppliers Group of Companies* (London, 1967).
Mitford, Mary Russell, *Our Village* (London, 1893, 1st edn 1824).
Mui, Hoh-Cheung & Mui, Lorna H., *Shops and Shopkeeping in Eighteenth-Century England* (Montreal, 1989).
Report from the Select Committee on the Rate of Agricultural Wages, and on the Condition and Morals of Labourers in that Employment, P.P. (392) (1824).
Report on the Administration and Practical Operation of the Poor Laws, B. Fellowes (1834).
Report on the Employment of Women and Children in Agriculture, Special Assistant Poor Law Commissioners, P.P. (510), xii (1843).
Rogers, James E. Thorold, *Six Centuries of Work and Wages. The History of English Labour*, 2 vols (London, 1884).

Rowntree, B. Seebohm & Kendall, May, *How the Labourer Lives. A Study of the Rural Labour Problem* (London, 1917, 1st edn 1913).
Samuel, Raphael (ed.), *Village Life and Labour* (London, 1975).
Shammas, Carole, 'The Eighteenth-Century English Diet and Economic Change'. *Explorations in Economic History* 21 (3) (1984).
Shammas, Carole, *The Pre-Industrial Consumer in England and America* (Oxford, 1990).
Sturt, George, *Change in the Village*, with Introduction by John Burnett (1984, 1st edn 1912).
Thompson, Flora, *Lark Rise to Candleford* (Harmondsworth, 1973).
Unwin, Mrs Cobden (ed.), *The Hungry Forties. Life under the Bread Tax* (London, 1905).
Westerfield, R. B., *Middlemen in English Business, particularly between 1660 and 1760* (New Haven, 1915).
Willan, T. S., *An Eighteenth-Century Shopkeeper. Abraham Dent of Kirkby Stephen* (New York, 1970).

DEREK J. ODDY

8 Formalizing relationships during the twentieth century

The roles of town and country in the British diet, 1860–1950

God made the country, and man made the town.
William Cowper (1731–1800), *The Task*.

Until 1914, the relationship between town and countryside was understood rather than stated. It was obvious to contemporaries that the countryside was engaged in agriculture, and that its larger estates provided opportunities for rural pastimes such as hunting and shooting which reflected past relationships between men of property and the land. However, the countryside was neither the centre of political power, as it had once been, nor was it any longer viewed as the means of feeding the town population. By 1860, the adoption of a free trade policy meant that the urban demand for cheap food had achieved political ascendancy over landed interests that sought to maintain the countryside as the towns' principal source of supply for commodities such as bread-grains, meat and vegetables.

The advocates of cheap food – the industrialists and urban middle classes – were eager to import foodstuffs from abroad if that was cheaper than producing food at home. Britain's trading capacities, both commercial and maritime, made this possible. During the second half of the nineteenth century imports of principal foodstuffs grew and from the 1870s agriculture in Britain began a decline that lasted for some 60 years. Cereal-producing areas of south-east England, particularly on heavy clay soils, went into decline from 1879 as grain prices fell. Under free trade, agriculture was subject to market forces but the government could not completely ignore its plight. The creation of the Board of Agriculture in 1889 was as far as the state went in unifying various countryside interests.[1] From 1885 to 1914, price falls extended into livestock and dairy produce.[2] By 1913, 53 per cent of cultivated land was under grass.[3] Few gave much thought to the effects of a free trade policy upon the food supply of Britain until the question was raised in the context of the growing militarism of the Edwardian years. In 1905, the Royal Commission on the Supply of Food and Raw Materials in Time of War[4] presented an assessment that showed the extent to which the sources of Britain's food supply had changed. Approximately 80 per cent of the grain for bread was imported and 45 per cent of meat came from abroad.[5] Even so, the strategic concept remained unchallenged: in time of war, the Royal Navy would protect Britain's food supplies; it would not be possible to starve Britain out of an European war.

The events of the 1914–1918 War proved to be different from expectations, since pre-war concepts had envisaged neither the length of the conflict nor the effectiveness of the submarine campaign against merchant shipping. For the first two years, supplies depended upon normal trade conditions and there were few, if any, signs of food scarcity. But the growing success of German submarines and the increasing demands upon British shipping forced change. By 1917, U-boats had sunk 2M tons of shipping yet one-third of Britain's ships was engaged in supplying the armed forces abroad, one-fifth was on loan to the Allies and only one-third was available for the home supply trade, including food.[6] The regulation of the food market – as far as it went – occurred in an ad hoc manner: the Home Secretary took control of the sugar trade successfully on 7 August 1914;[7] from October 1914 the War Office controlled imports of meat and a Royal Commission on Wheat Supplies was established in October 1916.[8] It was not until the beginning of 1917, that the government established a Food Production Department within the Board of Agriculture and Fisheries.

The Corn Production Act, 1917, brought price stability back to cereal farming with the ploughing up of grassland. It began a new policy of food production for strategic reasons which was continued after the end of the war by the Agriculture Act, 1920. The 1920 Act extended the policy of minimum prices for wheat and oats and improved the security of tenure for tenant farmers.[9] Food production expanded as County Agricultural Executive Committees began to bring land back under crops. With the rationing of food in 1918, the principle of tying consumers to retailers formalized the concept that urban consumers were completely dependent upon urban suppliers. Food came from shops; though the location of some ex-servicemen on horticultural smallholdings after the war through the auspices of the Land Settlement Association, revived the concept that townspeople in Britain were dependent on food producers in the countryside.[10] Nevertheless, by then, the old distinction between townsman and countryman was no longer relevant:

> Up to about 1914 there was still a rural England that was different, and always had been different, from the England of the towns: the First World War, with its call for a high-pressure technology and an all-embracing bureaucracy, accelerated the already evident process of modernization, and after about 1920 the rural mentality was found only among the older generation who had been unable to adapt to the new ways.[11]

The immediate postwar market suffered from the success of the wartime regime. Over-production of foodstuffs by primary-producers meant cheap imports seeking markets in the United Kingdom until world cereal prices collapsed. This led to the repeal of price support for cereal production in 1921.[12] However, economic thought was changing and the policy of free trade began to be questioned. An extensive review of Britain's food supply was carried out by the Linlithgow Committee in the early 1920s.[13] A temporary sugar-beet subsidy was introduced in 1925[14] and a system of distinguishing home-grown produce – the National Mark Scheme – was tried. The attempt by the Empire Marketing Board to stimulate milk consumption, in the hope of improving markets for dairy farmers was another attempt to stabilize agriculture. Nevertheless, by the late 1920s, British agriculture faced collapse.

The abandonment of free trade in 1931 brought significant changes. In 1932 agriculture and horticulture became protected and subsidized following the introduction of the Wheat Act and the Import Duties Act. The Agricultural Marketing Acts, 1931 and 1933 established Marketing Boards for hops, milk, potatoes, and bacon and pigs with the aim of raising prices for food producers.[15] This price-support system lasted, with some modifications, until 1973 when Britain accepted the Common Agricultural Policy upon entering the European Economic Community. In an endeavour to restrict the encroachment of the countryside by the spread of towns during the building boom of the 1930s, the concept of creating a green belt around major urban areas was introduced though, as suburban sprawl and ribbon development continued, the distinction between town and countryside became blurred, since more and more building plots meant less land available for food production. Some further stability to agriculture was given by the reorganization of the sugar industry in 1936, which extended the sugar subsidy in arable areas, while a Wheat Commission and a Cattle (later Livestock) Commission were introduced to handle other subsidies.[16] The attempted organization of the agricultural industry was in line with the corporatism of the interwar years and, however limited its achievements in improving farmers' incomes, provided some framework for emergency planning.

Unlike pre-1914 assessments, defence plans in the later 1930s assumed that in a future war much of Britain's food would be imported, but that there would be a major shortage of shipping in which to carry it. Any scarcities of food would bring about the inflation seen between 1914 and 1918, the prevention of which was a principal concern to the planners.[17] The stimulation of home food production was therefore essential. The Ministry of Agriculture acquired powers to control land use for cultivation and to direct food production by the revival of the system of County War Agricultural Executive Committees. Food production in the United Kingdom expanded by some 25 per cent above prewar levels by 1943/1944. By comparison with the First World War, the intensification of food production was dramatic:

> By 1918/1919, the peak of the effort resulting from the First World War, the net output of calories had been increased by about 24 per cent, compared with the years before the war; in the Second World War, the same increase had been achieved by the second harvest of the war and by 1943/1944 this had been quadrupled.[18]

With the passing of the Town and Country Planning Act, 1947, all land in Britain become zoned and the artificial division between town and country was complete. Once the Second World War was over, increased agricultural output became a central concern for British governments wishing to limit hard currency expenditure on food imports from North America. The postwar Labour government sought to increase output by 50 per cent over prewar levels and the Conservative administration of the early 1960s sought to raise output to 60 per cent of prewar figures.[19] The Agriculture Act, 1947, led to a period of stable prosperity for the countryside which lasted until 1973. Annual price reviews and guaranteed markets increased farmers' incomes, while the efficiency of farming was subject to the National Agricultural Advisory Service and to the continuation in peacetime of the system of County Agricultural Committees.[20] By effectively subsidizing production of all major crops in Britain, the governments of the 1950s and 1960s committed themselves to the last phase of Britain's cheap food policy before inflation in the 1970s, and entry into the Common Market drove food prices up.

Food processing and setting standards: creating the urban diet

From the mid-nineteenth century onwards, supplying food to consumers in the expanding towns created endless problems of transport, storage and distribution for wholesalers. As towns grew, the advance of bricks and mortar on to farmland made the retention of links with the countryside for towndwellers more and more difficult to maintain. Moreover, the local area of country around any British town provided very little of its food supply other than some seasonal fresh meat, poultry, eggs, fruit and vegetables. More significantly, the urban diet in the later nineteenth century came to depend to a considerable extent not so much on the countryside as on the staple items of the diet being processed on an industrial scale in the towns themselves. Thus millers and bakers, and brewers and distillers were among the first to apply mass production techniques to food. Margarine was another food industrially produced, though the shops that sold it tried to discount the artificiality of its origins by calling themselves "dairies". Otherwise, improvements in the variety and availability of foodstuffs for the urban housewife resulted more from the major retailing chains that grew up in the later years of the nineteenth century. None of them was national in coverage, though the Co-operative Wholesale Society, formed in 1863, became suppliers on a national scale to the various local retail co-operative societies. These commercial factors pointed to a decline in home-produced foods in Britain during the decades before 1914,[21] though the food market reflected regional and local differences, and seasonality of food supplies was still strong.

Technology to support the growing scale of the food supply system in the early twentieth century was limited. Refrigeration and cold storage were still new and the supply of artificial ice was insufficient for needs in high summer. Canning was still only used for a minority of foodstuffs. Traditional methods of food preservation by drying, smoking, salting or in sugar predominated. Although population growth began to slow down in the late nineteenth century, the population of the United Kingdom was still increasing at more than one per cent per annum up to 1911. In those circumstances, expanding the food supply and extending the shelf life of the more perishable items was important for wholesalers and retailers at all levels of the food market. The opportunities for adulteration and the use of additives and improvers was irresistible, and the early twentieth century experienced some notable instances where consumers' health was seriously affected as producers cut corners. In 1900, the use of glucose contaminated with arsenic in the brewing of beer in Lancashire and Staffordshire led to 6,000 cases of food poisoning, of whom 70 died.[22] The establishment of the Royal Commission on Arsenical Poisoning, 1903, focused public awareness on the low production standards of food and drink. Concern was heightened still further by the publication of Upton Sinclair's novel, *The Jungle*, in 1906.[23] In the years before the outbreak of war in 1914, the poor quality of town food in Britain was not universal: rather it had acquired a characteristic division based largely on income; the better-off bought fresh and locally-supplied food but amongst the working classes, the bulk of the food was cheaper imported food which, for technical reasons of preservation and marketing, was of questionable quality.

Imported wheat and poor quality flour caused problems for many bakers in towns. Wet, poorly risen flavourless loaves were common complaints, and in the warmer months, outbreaks of "rope" (*bacillus mesentericus*) – a bacterial infection of bread – occurred frequently up to 1914.[24] Biscuit factories, having successfully mechanized biscuit production in the mid-nineteenth century, began large scale production of slab cake in the 1890s, which the retail chains sold cheaply. By 1914, there were few areas of Britain where home-baking of household bread had survived[25] and bread, cakes and biscuits were predominantly "shop goods". A similar class division in commodity quality occurred in the meat trade: the frozen-meat trade expanded dramatically in the early years of the twentieth century. It became widely used in hotels, restaurants and was accepted by the Army: county councils bought it for poor law use and asylums. Servant-keeping families assumed they were eating home-killed meat, though they were largely dependent on their servants' honesty in dealing with tradesman.[26] By 1905, London and larger urban centres relied mostly on foreign meat supplied by the frozen-meat chains of meat retailers, though small towns, particularly in eastern England, obtained much of their meat from local sources.[27] Fish was another perishable commodity that suffered from the lack of refrigeration or even adequate supplies of ice. In consequence, much of the fish in urban fishmongers' shops early in the twentieth century was stale when sold and, with the conservative nature of the public's taste for fish, much "offal", i.e. the unsaleable varieties, only found its way into the food supply via the fish-and-chip shop.

In the late nineteenth century, poor quality butter, heavily salted and with added water, was common in towns[28] and this encouraged the production of butter substitutes. "Milk-blended butter", probably containing in excess of 25 per cent water, appeared as late as 1900 and was held by the courts not to be within the scope of the Margarine Act of 1887.[29] Margarine was poor quality stuff before the early years of the twentieth century when the hydrogenation of vegetable oils stabilized its consistency. But its use was expanding from 1909 onwards and margarine was an important commodity in the growth of retail chains, such as Maypole Dairies, and those that sold the products of the Dutch firms, Jurgens' and Van den Bergh's.[30] Nevertheless, imported butter supplies almost doubled in the years between 1890–1894 and 1910–1914,[31] and cheap butter from Denmark, Australia, Russia and New Zealand, limited the growth in margarine consumption. Milk was another commodity that was beginning to travel considerable distances to urban markets, in many cases from country areas hitherto devoted primarily to the making of butter and cheese. Within living memory of townsmen in 1914 most "milk drunk in the towns was produced either in the town itself or within a few miles of it".[32] But "town dairy" milksheds were in rapid decline and the spread of building was threatening many "suburban farms".[33] Wholesale contractors were buying milk from farms between 50 and 150 miles from London, yet cooling and refrigerated transport was rare in the liquid milk trade.[34] In consequence, much milk, if not actually adulterated by the addition of water, contained substances claimed to act as preservatives. As the market for liquid milk expanded, so the supply available for cheese making declined. Before 1914, cheese making in Britain remained largely a farmhouse industry disposing of its produce through cheese fairs. The demand from towns outstripped this supply, especially in view of the seasonal availability of milk. From the 1890s to 1914, imported cheeses amounted to around 2.3–2.4M hundredweights[35] per year. Canada supplied around one-half that amount in the form of factory-produced Cheddar-type cheese and New Zealand imports increased rapidly to around one-quarter of the total imports by 1910–1914.[36] Another commodity, both seasonal and perishable, that had only recently entered into the urban food market within the previous twenty years or so was the fresh hen's egg.[37] The bulk of the egg supply was in preserved eggs, which, said William Cobbett, were "things to run from not after".[38] Russian eggs were the cheapest but often the stalest "being almost putrid when sold in England, and chiefly used in manufactures, for which, at a low price, they answer perfectly".[39] Cold storage had no impact on the egg trade in Britain before 1918 and even in the 1920s, English farm eggs could be described as "dangerous stuff to deal with".[40]

Most vegetables were locally grown, though the range available was somewhat restricted and some imports, principally potatoes and onions from European suppliers, were necessary. Otherwise, vegetables were essentially seasonal: cabbage, sprouts and similar greens for the winter with carrots, onions and root vegetables from the late summer onwards. Fruit was also very seasonal, with soft fruit in the summer and hard fruit, such as apples and pears, in the autumn. Apples, bananas and oranges were part of a growing fruit trade before 1914, and were cheap enough to enter working-class diets occasionally. Prejudice against eating raw fruit persisted amongst the working classes until well into the twentieth century. For children, in particular, "summer diarrhoeas" were attributed to the "corrupting influence" of fruit.[41]

Fruit grown under glass, such as grapes and peaches were sold at prices that made them part of a luxury trade centred largely on London, where melons and pineapples were high-priced rarities.[42] The lack of cold storage in the fruit-growing industry made the whole trade highly perishable and kept prices extremely high.

Food quality and standards

Urban food supplies became increasingly subject to regulation in the early years of the twentieth century. The professional development of public analysts in the last quarter of the nineteenth century gave governments the confidence to investigate food production in a more systematic matter. The Sale of Food and Drugs Act had been passed in 1875[43] but later legislation in 1899 expected all areas of Britain to have Medical Officers of Health and Public Analysts in post. Their impact before 1914 was limited to trying to avoid the grosser forms of contamination of food and food premises and the nature of their work can be followed in the pages of *The British Food Journal*, founded in 1899. Its circulation made it almost an in-house journal for public analysts, which meant that principally it contained law reports of offences under the 1875 Sale of Food and Drugs Act, and the Act that replaced it in 1899. Other legislation, such as the Margarine Act, 1887, also provided case reports of interest to its readership. The provisions of new legislation were reported in full and the annual report of the Government Laboratory was also reproduced for readers' information. Between 1899 and 1914, *The British Food Journal*'s pages highlighted food legislation, adulteration, discussed the composition of foods and reported on production and preservation techniques, particularly the canning process, that led to new developments in the British diet. In summary, *The British Food Journal* provided a record of contemporary attempts to eliminate the imperfections of the urban supply system.

Public analysts continued to put pressure on retailers to reform their practices throughout the interwar period. The early legislation had been directed against the food retailer in British towns, particularly the small-scale trader. The proprietors of back-street and corner shops, market stallholders and retail dairymen found themselves before the magistrates for various offences, generally associated with attempts to extend the amount or marketable life of various commodities. In some trades, such as fishmongers and butchers, the traditional "open shop" – where food was displayed to passers-by without protection of windows – was still in use though, during the inter-war years, retail food marketing was being transformed by the chains of multiple retailers. They had mostly begun as provision merchants but, as they expanded the range of their products, were blurring the former distinction between the provision merchant and the grocer who sold dry goods. While the public analysts mounted a campaign against the open shop during the 1920s, the standardization and pre-packaging of commodities such as tea, margarine and soap was beginning to characterize the grocery trade. Branded goods were widely advertised in newspapers, on billboards, on railway stations, in theatres and cinemas. The application of food technology to the manufacturing of cakes, biscuits, breakfast cereals, pies, patent foods of varying kinds and soft drinks created products for the food retailer in which the manufacturer's price became more significant than the shopman's skills of cutting and weighing. Instead, for the small shopkeeper, his most important attribute came to be providing the variety of competing brands that met consumers' response to the food manufacturers' advertisements.

This meant that the quality of food became the concern of the factory bakers, biscuit manufacturers, confectioners, the suppliers of jams, pickles, sauces, relishes and condiments that appeared on the retailers' shelves. Food quality became associated with the names of the manufacturers of leading brands – Cadbury's, Rowntree's, Huntley and Palmer's, Tate and Lyle's – including also names like Heinz and Kellogg as American products arrived in the British market. It was the food manufacturer, therefore, rather than the individual retailer, who became involved with product standards. Once products were sold wrapped, packaged, or supplied in containers such as bottles or cans, the manufacturers had some degree of control over the quality of their product as it reached the consumer. For those supplying goods "loose" or unwrapped, such as bakers and butchers, quality was more difficult. Thus, by the 1930s, Associated Biscuit Manufacturers Ltd, United Biscuits Ltd, E H Nevill's, Garfield Weston, the British Sugar Corporation, Tate & Lyle Ltd, Unilever Ltd, and Vestey's were important suppliers to the retailers. In bread, biscuits, sugar, chocolate confectionery, margarine and frozen meat, the dominant position of a small number of firms had the effect of imposing rigidities upon the food market in much the same way as oligopolies developed generally in British industry during the interwar years.

Dietary change and nutritional status

It has generally been assumed that change towards modern patterns of food consumption began in Britain with a fall in the consumption of bread and flour during the late nineteenth century.[44] In *The Englishman's Food*, the Drummonds thought this decline:

> to be related to a rising standard of living for the falling curve representing bread and flour is complementary to the rising curve for sugar and sweetmeats. It is also related to a rise in the consumption of meat. These relationships reflect the fact that bread is the staple food of poverty and that people eat much less of it when they can afford to buy meat and to indulge in the type of dish with which sugar is eaten.[45]

Their views are undoubtedly correct in general terms. In Table 8.1, possible trends in food consumption are suggested between 1860 and 1914. The comparison of food consumption is based on Edward Smith's 1863 survey against means of the family budget surveys for the Edwardian period. The inverse relationship between bread and sugar is clear. In both in town and country bread consumption fell from over nine pounds (4 kgs) per week to around six pounds (2.75 kgs) per week. Sugar consumption, the other part of the relationship, more than doubled in both town and country over the same period. The difference between town and country diets occurred in fats and meat. While the amounts of fats and meat consumed was unchanged in the countryside, in towns the consumption of fats rose by 59 per cent, from 4.7 ounces (133 g) to eight ounces (227 g), and meat consumption increased by 61 per cent, from 0.8 pounds (360 g) to 1.3 pounds (590 g).

Table 8.1 *Weekly food consumption of working-class families shown by dietary surveys.*

Date	No. of families	Bread (lb) (kg)	Potatoes (lb) (kg)	Sugar (oz) (g)	Fats (oz) (g)	Meat (lb) (g)	Milk (pt) (l)
Urban							
1863	125	9.1 (4.10)	2.4 (1.09)	7.9 (224)	4.7 (133)	0.8 (363)	0.8 (0.45)
1902–13	2,031	6.7 (3.00)	2.9 (1.32)	15.7 (445)	8.0 (227)	1.3 (590)	1.7 (0.97)
Rural							
1863	377	11.6 (5.27)	4.4 (2.00)	6.6 (187)	5.2 (147)	0.9 (408)	1.6 (0.91)
1902–13	301	5.7 (2.59)	4.4 (2.00)	13.3 (377)	5.1 (145)	0.9 (408)	2.0 (1.14)

Sources: For 1863, see Barker, Oddy & Yudkin (1970), 43; for the Edwardian surveys, see Oddy (1971).

Evidence from these and other dietary surveys indicates that most of the change suggested by the Drummonds came within the period between the 1860s and the 1880s, with less change occurring from 1890 to 1914. Although the last quarter of the nineteenth century is regarded by historians as one of rising real wages, neither food supply estimates nor evidence of food consumption from working-class families point to any increase in meat consumption taking place between 1890 and 1914 in either town or country. Similarly, the increase in the amount of sugar consumed was much smaller after the 1880s. The fall in bread and flour consumption moved in parallel: in urban diets, it was down to 6.7 pounds (3 kgs) by the end of the 1880s, at which level it remained until the outbreak of the First World War.

Table 8.2 shows the effect of this change in nutrient terms. If bread consumption fell, energy intakes would decrease unless offset by increased amounts of other foods. Thus, in Table 8.2, the energy intake of rural diets fell from over 2,700 kcals (11,500 kj)/day to under 2,200 kcals (9,000 kj)/day as the fall in amounts of bread eaten outweighed the increase in sugar consumed. In urban areas, however, the fall in bread consumed was offset by more sugar, fats and meat, so that energy intakes rose from under 2,200 kcals (9,000 kj)/day to over 2,400 kcals (10,040 kj)/day. However, the impact of these changes varied according to the distribution of food within the family. In addition, the importance of the environment in the health of large sectors of the urban working class cannot be over-stated for the period before 1914. After the 1914–1918 War, its effect was sometimes questioned. M'Gonigle and Kirby, for example, noted that:

environment is not the only factor so operative, and possibly not the most important one. Most medical officers who served in France during the Great War were impressed by the healthiness of the troops who lived, often for long periods, under environmental conditions which, even at their best, could only be described as unsatisfactory.[46]

Table 8.2 *Daily nutrient intake of working-class families shown in dietary surveys.*

Date	No. of families	Energy value (kcal)	(kj)	Protein (g)	Fat (g)	CHO (g)	Iron (mg)	Calcium (g)
Urban								
1863	125	2,190	9,163	55	53	370	12.5	0.36
1902–13	2,031	2,431	10,171	73	67	377	12.2	0.47
Rural								
1863	377	2,760	11,545	70	54	460	15.9	0.48
1902–13	301	2,172	9,088	56	53	367	11.0	0.41

Sources: as Table 8.1.

While that might indeed have been true for men living on full Army rations, it would not be dissimilar from the conclusion reached by many social investigators before 1914 who commented favourably on the health and strength of the male labourer. The urban environment before 1914, however, was not only inhabited by adult males but was the background to birth and child-rearing. Although public health administration in the last five years or so before 1914 received a new impetus in the hands of medical officers like Newsholme at the Local Government Board and Newman at the Board of Education, both convinced environmentalists, the attack on the poor environment of inner urban districts could not change inadequate diets. Newman, writing in 1911 on the health of the schoolchild, expressed clearly the effect of the urban environment in a single sentence. "At the best, however, and in anything like the social and economic conditions which now prevail, the work which the Local Education Authority can do in directly preventing and remedying the evils of malnutrition has somewhat definite limits."[47]

Change in the twentieth century

Full employment during the First World War was instrumental in breaking the pattern of restricted diets that had been a threat to health and growth in the nineteenth century. But the collapse of the post-war economic boom in 1921 led to high levels of unemployment during the 1920s and 1930s, particularly in the older urban industrial areas. Not all the gains were lost after the War. Most change in the diet can be seen in the higher levels of sugar and fat consumption, and the growing purchases of packaged groceries that were becoming part of inter-war meals. Areas of poverty still existed, none the less, where diets, as restricted as any in the nineteenth century, were the norm. It was the effect of poverty that led John Boyd Orr to claim, much to the annoyance of the government and the officials of the Ministry of Health, that half the population had a less than adequate diet.[48]

There was a considerable variation in nutrient intake amongst the families surveyed by nutritionists in the 1930s. In the northern towns of Newcastle, Manchester, Stockton and Merseyside, and in Scotland and the Rhondda mining community, energy intakes were around 2,400 kcals (10,040 kj)/day or less. Although this showed an advance on nineteenth-century energy intakes in the range of 2,000–2,100 kcals (8,370–8,790 kj)/day, it seems that industrial areas of Britain had done no more than hold on to the pre-1914 energy intake levels. Other surveys showed advances in energy levels to over 2,500 kcals (10,460 kj)/day. All these surveys made in the 1930s indicate higher protein and fat intakes than was common in the nineteenth century but suggest that any expected rise in energy levels was largely offset because of falling carbohydrate intakes. With the exception of the diets collected by the Women's Co-operative Guild, the 1930s surveys show carbohydrate levels well below the 380g/day common in the nineteenth century and, by the 1930s, the proportional sources of the energy in the typical English diet were 12 per cent protein, 34 per cent fat and 54 per cent carbohydrate. Iron and calcium intakes were improving in the 1930s but in the northern towns in England, the calcium level – the target of the Empire Marketing Board's milk campaign – was still far too low for health and growth.

The Second World War and after

The application of food rationing during the Second World War has generally been regarded as improving nutritional standards in Britain, though despite the introduction of a dietary survey in 1940, the data available were never sufficient to monitor the nutritional status of the population fully. Since prewar Britain had consisted of one-third well-fed, one-third adequately fed and one-third whose diet was deficient, rationing raised the nutritional status of this last group to the level of adequacy. Despite the rise in working-class incomes during the war, there were still poor people with inadequate diets; those who spent least on food were still dependent on "bread and spread" rather than cooked meals.[49] However, from a longer-term view of dietary change, the enforced restrictions on food choice during the 1940s and early 1950s inhibited change. The British diet shown in tables 8.3 and 8.4 was remarkably stable. What the tables also show is that economic policy rather than nutritional status was of major significance in determining consumption patterns. For example, the greatest restrictions on the British diet were not caused by wartime attacks on Britain's food supplies but rather in the postwar years, when American aid to Britain ceased. The "dollar gap" – the shortage of hard currency with which to buy North American imports such as wheat – became most acute in 1947, so that for two years after the war – 1946 and 1947 – energy intakes were lower than during the War itself. The effect could be seen between 1945 and 1947 from the evidence of a setback in growth rates, both in heights and weights, affecting children at all ages between five and 15 years.[50] This suggested that the nutritional status of the population was much closer to the limits of adequacy than has generally been assumed to be the case and that children's nutrition was below the optimum for growth. The rationing scheme had not eliminated the social gradient of heights and weights.[51]

Table 8.3 *Weekly food consumption, 1937–1950, per head of the population.*

Date	Bread[1)] (lb) (kg)	Potatoes (lb) (kg)	Sugar (oz) (g)	Fats (oz) (g)	Meat (lb) (g)	Milk (pt) (l)
1937	3.8(1.73)	3.2 (1.45)	17.3 (490)	11.6 (329)	1.4 (635)	2.8 (1.59)
1941	5.4 (2.45)	4.3 (1.95)	8.6 (244)	8.3 (235)	1.8 (816)	3.6 (2.05)
1942	5.1 (2.32)	4.3 (1.95)	8.4 (238)	8.7 (247)	1.6 (726)	3.8 (2.16)
1943	5.1 (2.32)	4.5 (2.05)	8.7 (247)	8.7 (247)	1.6 (726)	4.3 (2.44)
1944	5.2 (2.36)	4.5 (2.05)	9.0 (255)	9.2 (261)	1.8 (816)	4.4 (2.50)
1945	5.3 (2.41)	4.3 (1.95)	9.1 (258)	8.6 (244)	1.6 (726)	4.4 (2.50)
1946[2)]	4.3 (1.95)	4.4 (2.00)	15.0 (425)	8.2 (232)	1.7 (771)	4.0 (2.27)
1947	4.3 (1.95)	4.1 (1.86)	15.7 (445)	7.8 (221)	1.5 (680)	3.9 (2.22)
1948	4.7 (2.14)	3.7 (1.68)	16.4 (465)	8.9 (252)	1.5 (680)	4.0 (2.27)
1949	4.5 (2.05)	4.1 (1.86)	17.1 (485)	10.6 (301)	1.4 (635)	4.4 (2.50)
1950	4.2 (1.91)	4.0 (1.82)	15.9 (451)	11.5 (326)	1.9 (862)	4.6 (2.61)
1950[3)]	4.3 (1.95)	3.9 (1.77)	16.4 (465)	11.6 (329)	1.9 (862)	4.8 (2.73)

Sources: 1937 = calculated mean of English families (1046) in the Carnegie Survey; for 1940s, see *Report of the National Food Survey Committee,* 1951, 1952.

Note: [1)] Includes flour and other cereals.
[2)] 1946–1950 figures are for urban working-class households only.
[3)] All households.

Table 8.4 *Daily nutrient intake, 1937–1950, per head of the population.*

Date	Energy value (kcal)	(kj)	Protein (g)	Fat (g)	CHO (g)	Iron (mg)	Calcium (g)
1937	2,540	10,627	76	96	343	13.7	0.65
1941	2,340	9,791	73	n.k.	n.a.	12.4	0.61
1942	2,270	9,498	74	n.k.	n.a.	13.5	0.67
1943	2,270	9,498	73	86	n.a.	13.3	0.86
1944	2,390	10,000	73	94	n.a.	13.5	0.87
1945	2,375	9,937	76	92	n.a.	12.7	0.88
1946[2)]	2,305	9,644	78	86	n.a.	14.4	0.91
1947	2,310	9,665	77	82	n.a.	14.3	1.00
1948	2,390	10,000	77	88	n.a.	14.2	1.01
1949	2,425	10,146	76	95	n.a.	13.6	1.03
1950	2,440	10,209	77	102	n.a.	13.5	1.04
1950[3)]	2,475	10,355	78	101	n.a.	13.6	1.06

Sources and notes: as Table 8.3.

However, by the beginning of the 1950s, as food rationing began to be eased, Dr Magnus Pyke reviewed the food of the overwhelming majority of the British people in a book entitled *Townsman's Food*.[52] It may well be that we should regard his title as confirming that distinctions between town and country diets had largely disappeared. After ten years of rationing the townsman – or more particularly, the urban housewife – was thinking wistfully of times of former plenty. The concept of "fresh country produce" was an important part of this yearning, for it symbolized good quality food to the urban population of mid-twentieth century Britain. However, Magnus Pyke was uncompromisingly dismissive of nostalgia and feelings of "romantic regret for the simple foods of earlier centuries". From his pragmatic viewpoint as food scientist, the urban housewife with a tin-opener may have been a figure of fun for cartoonists in the 1940s, but her search for variety in the diet meant that few could live without canned meat, fish or fruit.[53]

Nearly fifty years later, the consumption patterns Pyke describes sound very strange to consumers today. As a guide to the diet of 1950, Magnus Pyke's book concerned itself largely with a description of the sources of food materials, i.e. bread and baked goods, vegetables, meat, dairy produce, etc., rather than processed food products. It was published in an age when confidence in science was much higher than today and Magnus Pyke, whose popular reputation rested upon his creed of bringing science to the people, saw the food industry as providing a 'social service'; after all, he wrote: "the food industry has to keep the community alive and in tolerable health".[54] There is little on food processing or restrictions on the use of materials by the Board of Trade, because the presentation and marketing of food is almost as absent from Pyke's book as food itself was from the Hammonds' description of town life in the nineteenth century.[55] In the 1940s, wrapped bread called for little by way of packaging or presentation. True, technology was important: Magnus Pyke referred to gas storage of fruit, and the use of sulphur dioxide, but when he claimed: "Today, foods 'in season' have been abolished. In the modern world, we expect all foods always to be 'in season'", we know that the effervescent optimism which endeared Pyke to radio and television audiences has carried him away.[56]

Commercial food processing was not without problems. Freezing ruptured the cellular structure of food, quality deteriorated; there was chemical change in fat and food could suffer from "freezer burn". Chilling was still found to affect flavour adversely. Pyke's view of processed food was typified by his description of ice cream, which, as an unrationed food, had an aura of luxury and excitement at the end of the 1940s:

> Modern British ice cream, therefore, is in general, a wholesome food; but its nutritional value may vary greatly in proportion with, firstly, the amount of dried milk it contains and whether the dried milk is from full-cream milk or from skimmed milk, Secondly, it may contain much or little fat. If ice cream is poor in milk and poor in fat, it must contain some other ingredient. This other ingredient may be flour or another cereal product, in which case the nutritional value of the ice cream will more closely resemble custard-powder custard than ice cream (using the name literally).

> On the other hand, the alternative ingredient may be water. If it is, the ice cream will gradually sink below the category of a food, and will have to take its place among such dietetic anomalies as jelly, of which a plateful contains a pinch of solid matter.[57]

It is tempting to think, there you have it. Excitement in British towns around 1950 was eating jelly and ice cream. There were processed peas, of course, dyed green and sterilized, but the new frozen vegetables were "a luxury".[58] In fact, the war had brought a marked deterioration in the quality and availability of some foods. Canned meat was of variable repute. The minimum meat content of processed meats such as "luncheon sausage, breakfast sausage, meat galantine and polony" had been reduced to 30 per cent. During wartime, food-poisoning rose; between 1941 and 1948, 47 per cent of all cases resulted from made-up meat products.[59] The quality of fish was a problem for inland towns, where, thought Pyke, the fish-and-chip industry, "might almost be said to have been created to overcome the poor quality fish caught by the modern fishing industry" and consumer resistance to wartime salt cod and post-war *snoek* had not wavered.[60] Cheese quality had deteriorated and the variety of cheeses available had diminished markedly. This was less a result of wartime circumstances than the pre-war changes in the liquid milk market during the 1930s. But rationing had accentuated the trend: if social justice was to be maintained, any commodity must be acceptable to the whole population: "rationed cheese must not possess too pronounced a flavour". This ruled out Stilton and Gorgonzola and reduced production to "Cheddar type" cheese.[61] The most dramatic effect on the food supply during the war, however, had been the almost complete removal of poultry from the diet on the grounds that they ate imported grain. In consequence, feed allocations to livestock farmers reflected government priorities, firstly, to feed cows for milk production, secondly, to feed sheep and beef cattle and lastly, pigs and poultry. By the middle of the twentieth century, therefore: "poultry has become a rarity on the urban dining table: its price puts it beyond the reach of the ordinary family – except on special occasions like Christmas Day".[62] That marks the end of an era: no greater contrast can be drawn between the urban diet before intensive farming and the present-day urban diet based on the modern supermarket, its freezer cabinets stacked with chicken dishes of all kinds, shapes and sizes, and its rotisserie ovens continually turning out ready-cooked poultry. In the high streets, the fast-food outlets sell chicken-burgers and chicken nuggets at all hours of the day.

In 1950, the urban diet in Britain left people longing for variety and freedom of choice. They were far from understanding the retailing revolution in foodstuffs that was to come, but no view of 1984 put before them by George Orwell[63] could have prepared them for the social change of the last fifty years. The recent dramatic advances in food processing, packaging and presentation and the contemporary growth of eating out were unimaginable in the Britain of 1950. Today, the consumer buys chilled or frozen ready-made dishes or complete processed meals, from a huge range of branded products, many of which originate in foreign cuisines. For the housewife in 1950 suffering under the restrictions of rationing, it would have been beyond belief to imagine that at the end of the twentieth century (rather than 1984), she would heat up in a microwave oven what she had brought home from the shop, while cooking, as such, had been reduced to a form of television entertainment. It would also have been impossible for the housewife of 1950 to imagine that at the end of the century she would fear the countryside as a source of danger. The countryside, for so long the symbol of good quality food, was now the source of foods associated with obesity and heart disease – cream, full-fat milk, cheese and meat eschewed by modern 'foodies' – and, in addition, the source of chickens and eggs infected with Salmonella, unpasteurized milk products contaminated by Listeria, and cattle infected with Bovine Spongiform Encephalopathy (BSE). Neither George Orwell nor even Magnus Pyke could have foreseen that!

NOTES

This chapter draws on material published in Derek J. Oddy, *From Plain Fare to Fusion Food British Diet from the 1890s to the 1990s*, Boydell and Brewer, Woodbridge, Suffolk, and Rochester, N.Y., 2003.

1. Winnifrith (1962), 23–24. The Board of Agriculture for Scotland was created by the Small Landholders (Scotland) Act, 1911 and, according to *The British Imperial Calendar*, began to operate in 1913.
2. Perren (1973), 109.
3. Pollard (1969), 8.
4. *Report of the Royal Commission on the Supply of Food and Raw Materials in Time of War*, 1905 (Cd.2643).
5. Court (1965), 189–190.
6. Marrack (1946), 164.
7. *First (Interim) Report of the Royal Commission on Sugar Supply*, P.P. 1917–1918 (Cd.8728) XVIII, 635.
8. Marrack (1946), 164–165; Pollard (1969), 43
9. Pollard (1969), 134.
10. One of the effects of the 1905 Royal Commission on the Supply of Food in Time of War had been that County Councils were required to make land available for smallholdings. By 1914 there were only 14,000 in existence. See Ashworth (1960), 64. In Scotland, the Board of Agriculture created a small number of crofts for ex-servicemen in the 1920s.
11. Wain (1966), xvii.
12. Pollard (1969), 135.
13. *Final Report of the Departmental Committee on the Distribution and Prices of Agricultural Produce*, 1924 (Cmd.2008).
14. British Sugar Subsidy Act, 1925.
15. Pollard (1969), 139–141.
16. Winnifrith (1962), 25–26.
17. Hammond (1951), 8.
18. Murray (1955), 243–244, quoted in Winnifrith (1962), 29.
19. Winnifrith (1962), 30.
20. Pollard (1969), 384.
21. Oddy (unpublished Ph.D. 1971), 80–83.
22. Pyke (1952), 188.

23. See *The Grocers' Journal*, 28 July 1906, 11–12, for the effect upon the trade in tinned meat.
24. *Baker and Confectioner*, 15 January 1909, 47; *Baker's Quarterly Magazine*, June-September 1906, 70.
25. *Baker and Confectioner*, 26 June 1914, 569.
26. *Report of the Select Committee on Marking of Foreign Meat*, P.P.1893–4, XII; Evidence: Questions 2825–2836.
27. *Report of the Departmental Committee on Combinations in the Meat Trade*, P.P. 1909, (Cd.4643) XV, para.7; see also the *Report of the Select Committee on Marking Foreign Meat*, P.P.1893–1894, XII.
28. *Report from the Select Committee on Food Products Adulteration*, 1894–1896, P.P. 1896, IX.
29. *Grocery*, February 1902, 106.
30. See Wilson (1954), ch.VII. By 1912, Jurgens' supplied Home and Colonial Stores and several small regional chains, while Van den Burgh's supplied Lipton's, Pearks', and Meadow's (including Keeloma). Maypole's margarine was supplied principally by Otto Monsted, a Danish company with a factory at Southall, Middlesex.
31. Cohen, (1936), 29.
32. Ibid., 17.
33. *Report of the Travelling Commission of Enquiry into the cost of Production of Milk*, P.P. 1919 (Cmd.233) XXV, 435, para. 14.
34. *Report of the Departmental Committee on Preservatives and Colouring Matter in Food*, P.P. 1901 (Cd.833) XXXIV, QQ. 3951–3953.
35. Hundredweight = 112 pounds (or approximately 50 kgs).
36. Cohen (1936), 36.
37. Wright (1902), 166.
38. Quoted in Brown (1900), 168.
39. *Encyclopaedia Britannica* (1910–1911), 220.
40. Ministry of Agriculture and Fisheries Economic Series, no. 10, *Report on Egg Marketing in England and Wales* 1926, 22.
41. Drummond & Wilbraham (1957), 379.
42. *Report of the Departmental Committee on the Fruit Industry of Great Britain*, P.P. 1905 (Cd.2589) XX, para. 68; also *Fruit, Flower and Vegetable Trades' Journal*, 13 December 1913, 955–956.
43. Earlier legislation – the Adulteration of Foods Act, 1860 and the Adulteration of Food, Drink, and Drugs Act, 1872 – was permissive and thus limited in its effectiveness.
44. Drummond & Wilbraham (1957), 299.
45. Ibid.
46. M'Gonigle & Kirby (1936), 143–149.
47. Board of Education, *Annual Report of the Chief Medical Officer*, 1910, para. 314.
48. Orr (1937), 44 (and Foreword to the First Edition). Orr stated that "a diet completely adequate for health, according to modern standards, is reached at an income level above that of 50 per cent of the population". He left others to deduce that the diet of the other 50 per cent was inadequate.
49. BMA, *Report of the Committee on Nutrition* (1950), para. 111, 40.
50. Ibid., paras. 184–191, 79–82.
51. Ministry of Health, *Public Health during Six Years of War* (1946), 121.
52. At the Censuses in 1951, 80 per cent of the population were classed as urban and 20 per cent as rural.
53. Pyke (1952), vii.
54. Ibid., 11.
55. Hammond & Hammond (1947, 1st edn 1934) does not discuss diet; water supply in towns is referred to in Chapter V – but only in respect of washing and town cleansing; 66–71. The Hammonds seemed more concern at the poor facilities for clothes-washing and the availability of hot baths than the need for drinking water. In *The Town Labourer*

(1978, 1st edn 1917), similarly the Hammonds themselves have nothing to say on food supply or diet beyond quoting the *Pioneer or Trade Union Magazine* of 19 October 1833: "We eat the worst food, drink the worst drink", 31, or noting that the payment of wages for gangwork or subcontracted labour (including the children who were the cotton piecers) took place in public houses. This was seen as a cause of early drinking; 34. Only in *The Village Labourer* (1978, 1st edn 1911) was there any discussion of diet and that, principally, through comments on the surveys of Davies and Eden. "Diet reform", the first section of chapter 6 is primarily a discussion of middle-class remedies for the reduction of expenditure by the labouring classes in the face of the rising prices of the 1790s. It is followed by a section on the introduction of the Speenhamland system.

56. Pyke (1952), 127.
57. Ibid., 125.
58. Ibid., 174.
59. Ibid., 63, citing the Ministry of Food Statutory Instrument No. 1509 of 1948. Also, 196.
60. Ibid., 67.
61. Ibid., 126–128.
62. Ibid., 81–84. He offers calculations derived from the School of Agriculture, Cambridge and the Imperial Bureau of Animal Nutrition (Rowett Research Institute), Aberdeen, that from 100 pounds of digestible feed, the dairy cow produced 35 pounds of animal protein, the pig, 20 pounds, the bullock and sheep, 8–15 pounds, whereas hens laid 30 pounds eggs and produced 20 pounds meat.
63. Orwell (1948).

BIBLIOGRAPHY

Ashworth, W., *An Economic History of England 1870–1939* (London, 1960).

Baker and Confectioner 1909, 1914.

Baker's Quarterly Magazine 1906.

Barker, T. C., Oddy, D. J. & Yudkin, J., *The Dietary Surveys of Dr Edward Smith 1862–1863* (London, 1970).

Board of Education, *Annual Report of the Chief Medical Officer*, 1910.

British Medical Association (BMA), *Report of the Committee on Nutrition* (London, 1950).

Brown, E, 'The British Egg Supply'. *Journal of the Royal Agricultural Society of England* LXI (1900).

Cohen, R L, *The History of Milk Prices* (Oxford, 1936).

Court, W H B, *British Economic History 1870–1914* (Cambridge, 1965).

Drummond, J. C. & Wilbraham, A., *The Englishman's Food*, rev. edn Dorothy Hollingsworth (ed.) (London, 1957).

Encyclopaedia Britannica, Eleventh Edition (Cambridge, 1910–1911).

Fruit, Flower and Vegetable Trades' Journal 1913.

The Grocers' Journal 1906.

Grocery 1902.

Hammond, J. L. & Hammond, Barbara, *The Bleak Age*, rev. edn (London, 1947, 1st edn 1934).

Hammond, J. L. & Hammond, Barbara, *The Town Labourer 1760–1832. The new civilization*, rev. edn (London 1978, 1st edn 1917).

Hammond, J. L. & Hammond, Barbara, *The Village Labourer*, rev. edn (London, 1978, 1st edn 1911).

Hammond, R. J., *Food Volume I The Growth of Policy*, History of the Second World War, United Kingdom Civil Series (London, 1951).

M'Gonigle, G. C. M. & Kirby, J., *Poverty and Public Health* (London, 1936).

Marrack, J. R., *Food and Planning* (London, 1946).

Ministry of Agriculture and Fisheries Economic Series, No. 10, *Report on Egg Marketing*

in England and Wales, 1926.
Ministry of Agriculture, Fisheries and Food, *Report of the National Food Survey Committee,* 1951, 1952.
Ministry of Health, *On the State of the Public Health during Six Years of War, Report of the Chief Medical Officer of the Ministry of Health 1939–1945* (London, 1946).
Murray, Sir Keith, *Agriculture*, History of the Second World War, United Kingdom Civil Series (London, 1955).
Oddy, D. J., *The Working-class Diet 1886–1914* (unpublished University of London Ph.D., 1971).
Orr, J. B., *Food, Health and Income*, 2nd edn (London, 1937).
Orwell, G. (Eric Blair), *Nineteen Eighty-Four: A novel* (London, 1949).
Perren, R., 'The Landlord and Agricultural Transformation 1870–1900', in: P. J. Perry (ed.), *British Agriculture 1875–1914* (London, 1973).
Pollard, S., *The development of the British Economy 1914–1967*, 2nd edn (London, 1969).
Pyke, M., *Townsman's Food* (London, 1952).
Report of the Departmental Committee on Combinations in the Meat Trade, P.P. 1909, (Cd.4643) XV.
Final Report of the Departmental Committee on the Distribution and Prices of Agricultural Produce, 1924 (Cmd.2008).
Report of the Departmental Committee on the Fruit Industry of Great Britain, P.P. 1905 (Cd.2589) XX.
Report of the Departmental Committee on Preservatives and Colouring Matter in Food, P.P. 1901 (Cd.833) XXXIV.
First (Interim) Report of the Royal Commission on Sugar Supply, P.P. 1917–1918 (Cd.8728) XVIII.
Report of the Royal Commission on the Supply of Food and Raw Materials in Time of War (Cd.2643), 1905.
Report of the Select Committee on Food Products Adulteration, 1894–1896, P.P. 1896, IX.
Report of the Select Committee on Marking of Foreign Meat, P.P.1893–1894, XII.
Report of the Travelling Commission of Enquiry into the cost of Production of Milk, P.P. 1919 (Cmd.233) XXV.
Rowett Research Institute, *Family, Diet and Health in Pre-War Britain* (Dunfermline, 1955).
Wain, John (ed.), *Selected Stories of Thomas Hardy* (London, 1966).
Winnifrith, Sir John, *The Ministry of Agriculture, Fisheries and Food* (London, 1962).
Wilson, C., *The History of Unilever*, 2 vols (London, 1954).
Wright, L., *The New Book of Poultry* (London, 1902).

ANNEMARIE DE KNECHT-VAN EEKELEN

9 From food to leisure

The changing position of the *jardin ouvriers* movement in the Netherlands from 1880 to the present day

Allotment gardens have existed all over Western Europe since the late nineteenth century. Their history is not only a history of the changing relations between men and environment, but also a history of emancipation of the working classes. In the Netherlands the first allotments were created in rural areas in the first half of the nineteenth century. It took until the 1890s before the potential of these gardens to improve the economic situation of the labourers attracted attention. Major initiatives were taken by the *Maatschappij tot Nut van 't Algemeen* (Society for Public Welfare), a Protestant charity with its roots in the Enlightenment. As a result of the first international congress on *jardins ouvriers* in Paris, in 1903, the Dutch branch of the Roman Catholic Society of St Vincent de Paul became more interested in allotments. At the same time interest grew amongst the labour movement and the councils of larger cities. The urban garden movement originated in Amsterdam in 1910. Urban garden societies were organized differently and had other objectives than the rural allotment societies.

After the First World War urban gardens lost their association with charity. The tenants undertook the management of the gardens themselves. They started using their gardens not only for growing crops but for recreation as well. This development went on after the Second World War so much so that in the 1970s the majority of urban gardens were used for recreation only. The summerhouses built in the gardens constituted relatively cheap second homes for those who lived in the narrow streets of the inner cities with little air and no facilities for outdoor life. Even today other recreational activities can not compete with these urban gardens in terms of the low costs and nearness to people's homes. Urban gardeners value their work in the garden, but they enjoy the time being spent there as well. In the following sections, the changing position of the *jardin ouvriers* will be examined.

Charity and allotments

In 1881 Jan Bruinwold Riedel (1849–1921), a Dutch theologian, was assigned to the post of country parson in a small village in the far north of the province of Groningen. There he had to manage the so-called *pastorie-landen*, the grounds belonging to the church. Contrary to the ordinary procedure of farming out these lands to the local farmers, he let small parcels to the village labourers. These people cultivated their plots, growing vegetables and potatoes for their own use. This was Bruinwold Riedel's first experiment with what he called "providing allotments for the 'homely' folk". Six years later he was assigned to the post of general secretary of one of the leading Dutch charities, the *Maatschappij tot Nut van 't Algemeen* (*Het Nut*), that took a major interest in the education of working-class people and the improvement of their living conditions.

The still active and well-known organization *Het Nut* had been established in the eighteenth century (1784) by a Mennonite parson, Jan Nieuwenhuijzen. *Het Nut* was organized as a society consisting of more or less independent branches; in 1885 the number of branches reached a maximum of 343; it took until 1957 before this number was reached again. The education of the lower classes was the main aim of this bourgeois enlightenment initiative. They wanted to reach this goal by the improvement of teaching and the publication of simple booklets. *Nut*'s branches were established in the Protestant parts of the country. They created schools, libraries, and from 1818 onwards also savings banks to support the virtues of thrift and carefulness. As Bruinwold Riedel served *Het Nut* for more than twenty-five years (1887–1915) in the prominent position of general secretary, he had every opportunity to disseminate his ideas on people's welfare and education. He was a prolific writer who produced important studies for *Het Nut* as an organization, as well as popular booklets for the *Nut*'s members. One of his many activities was the establishment of an allotment movement in the Netherlands.[1]

At the end of the nineteenth century the main goal of all initiatives concerning allotments was the welfare of the labourers. Several organizations occupied themselves with this task. There was the Protestant church, the parochial relief and the city's poor relief board, but also the *Algemeen Nederlandsch Werkliedenverbond* (Dutch Workers' League), the first political party for labourers established in 1886, and Patrimonium, another organization of workers. These labourers' societies also took an interest in cultivating allotments as a means of contributing to the family income.[2] All their activities concerned people who worked in the cities as well as those in the country, as none of them had access to land of their own. However, it took until 1903 before a major breakthrough of the idea of the *jardins ouvriers*, as they were often called in the Netherlands, took place. The first international congress on *jardins ouvriers* in Paris in that year caught the public eye, so Bruinwold Riedel, who of course attended the meeting, took his chance and published an overview of the national and international achievements of workmen's gardens. His book, *Arbeiderstuinen in binnen- en buitenland* (Workmen's gardens at home and abroad), is my main source on the early developments of the *jardins ouvriers* in the Netherlands.

It was not Bruinwold Riedel's first publication on the topic. In 1893 he had produced an overview in the *Nut*'s yearbook and in 1896 he had published a booklet in one of the *Nut*'s series.[3] But already before Bruinwold Riedel joined the *Nut*'s organization, some of the *Nut*'s branches had experimented with allotments that were rented to labourers for a small sum. The first initiative dated back to 1838 when the *Nut*'s savings bank in Franeker, one of the smaller towns in Friesland, let out pieces of land they owned. As one of the important activities of *Het Nut* was the establishment of its own saving-banks, the *Nut*'s branches had ample opportunity to buy land and farm it out on their own conditions. In the province of Groningen especially, where large landowners predominated, there was no chance for labourers to get access to property. Several of *Nut*'s branches in Groningen, the first as early as 1847, followed the example of Franeker. In 1904 Bruinwold Riedel counted a total of fifteen branches that owned land and rented it as allotments.

Table 9.1 *Number of allotments established in the nineteenth century.*

Year of establishment	Place	Number of allotments in 1904
Nut		
1838	Franeker	209
1847	Uithuizen I	344
1847–1848	Uithuizen II	ca. 200
1862	Loppersum	ca. 80
1854	Oude Pekela	216
1872	Wildervank	82
1880–1886; 1899	Veendam	20
n.a.	Holwerd	32
1886	Bedum	36
1892	Onderdendam	40
1893	Winsum	18
1894	't Zand	39
1894	Joure	ca. 76
1894	Noord-Beveland	39
1895	Heveskes	56
Other organizations[1)]		
1886	't Bildt (local council)	7
1890	Oostwoud (protestant church)	52
1893	Arnhem (tenants' society)	15
1896	Goes (poor relief council)	56

Source: Bruinwold Riedel (1905).

Note: [1)] In addition to the above mentioned organizations also *Broek in Waterland* (poor relief council) established allotment(s). However, neither information on the year of establishment, nor the number of allotments are available.

Only "well-behaved" labourers, widows and other respectable but poor citizens were deemed suitable by *Het Nut* to rent an allotment. They were supposed to grow potatoes, vegetables, beans, peas, etc. for their own use. In Bruinwold Riedel's view the *arbeiderstuin* (workmen's garden) added to the moral uplift of the workman: it prevented poverty, increased prosperity and self-esteem and it contributed to the strengthening of family ties. Each workman should have a well-equipped house to live in and a garden to work in.[4]

Allotments and poor relief

Het Nut was not the only charity that took an interest in the gardens. In Friesland, where the situation was comparable to that in Groningen, a *Vereeniging 'door arbeid tot verbetering'* (Society for Betterment through Labour) was established with the aim "to diminish the social distress in Friesland".[5] An inquiry by this Society showed that around 1900 almost 40 different organizations in Friesland let allotments to labourers. Among them were Protestant churches, various societies, poor relief funds, the local council, labour organizations, private persons, an orphanage, a savings bank and even a health insurance fund. A total of 1,284 gardens covering over 112 hectares were rented out. The Society for Betterment through Labour itself owned 106 gardens in six villages. The purpose of these gardens was to give labourers the opportunity to grow crops for their winter provisions. In Friesland it showed that fewer people had to make use of the poor relief funds during wintertime because of these extra supplies. Several examples were given from the islands of Ameland and Schiermonnikoog where in contrast to the other parts of Friesland all labourers had their own gardens, especially for growing potatoes. In these islands the poor relief fund paid only a few hundred guilders compared to a total of around Dfl 688,000 in the whole province. Moreover, in this way a new generation was created, no longer a generation of paupers, but one with experience in horticulture that could be used in their future lives. And last but not least: the poor relief costs had gone down. The national government shared these views and published a report stating the positive effects of allotments on the social situation of the lower classes.[6] The Minister of Internal Affairs asked for an experiment to prove the idea that allotments prevented people from falling into poverty.[7]

It is interesting to note that in these early initiatives in the Netherlands any input by the Roman Catholic organizations was lacking. The Dutch branch of the Society of St Vincent de Paul first mentioned the *jardins ouvriers* in one of their meetings in 1901, even though the society had been very active in France and Belgium for some considerable time. They stated that gardening "with Christian spirit" would increase the economic and moral status of families. An experiment was started in The Hague on a piece of the railway company's land. It appeared that the Society of St Vincent de Paul found that only seven families out of the 600 their charity supported were suitable to be garden-holders.

Gardens along the railway in the Amsterdam urban garden complex Nut en Genoegen *(ca. 1975). Because of the reconstruction of this railway 150 gardens in the complex had to move elsewhere.* Source: *Baar (1997), 118.*

This shows one of the obstacles in the cities in creating workmen's gardens for paupers who had no experience in gardening. But there are other reasons as well why the development of the allotments was slow. Different arguments were used to show the negative effects of gardening. Some said that labourers with a garden were content with a lower salary, because they had their winter provisions; so it seemed not to be in the interest of the labour unions. Others mentioned that the wife and children had to work in the gardens because the labourer himself had too little spare time. To get a better picture of the situation the Dutch Workers' League organized an inquiry in 1901. They concluded that labourers should be able to acquire or rent land of their own; city councils had to make this possible and labourers had to unite in pressure groups to demand their rights.[8]

One of the labour unions that asked for this right was a Union of Railwaymen. In 1901 they were the first to organize a mutual society to exploit pieces of land along the railway from Utrecht to Hilversum. Only railwaymen could become members. This type of organization was successful: in 1955 the allotments along railways covered an area of 259 hectares.[9]

The first international congress on jardins ouvriers

Bruinwold Riedel took the opportunity to publish data from the report of the first international congress on *jardins ouvriers* held in Paris in 1903. He summarized the congress report and translated French regulations, noted Belgian garden layouts and compared the Dutch situation with those in other countries. It had been a large meeting: around 700 participants are listed in the report of the congress; among them were three people from the Netherlands: L. van Aken, J. Bruinwold Riedel and M. P. de Clercq. The objectives of this meeting were: "To consider all aspects of the moral improvement to be achieved by working the soil; we know this as *jardins ouvriers*".[10] Over two days the conference discussed a variety of topics, including the statistics and organization of gardens, their regulations, the moral improvements, material results, and gardens and hygiene. An extra day was devoted to the "means for development and propaganda".[11] To obtain information on these topics a questionnaire had been used and in his book Bruinwold Riedel presented the data he had collected to report to the congress.

France appeared to play a leading role in the development of this type of garden. At the congress a list of 294 French groups of gardens was presented covering 6,453 garden plots with a total surface of more than 269 hectares. Important points for discussion were:

- who owned the gardens?
- was the garden for free or was there a rent?
- what were the most profitable products?
- should one grow flowers?
- were the products eaten by the family only or were they sold as well?
- were there contests for the best kept garden?
- was there a garden shed?

But moral questions were also addressed: did gardening diminish alcoholism, tuberculosis and infant mortality? The purpose of the garden was "to put the worker from the factories and workshops in touch with the soil and the sun again, and to give him a share in sunlight, fresh air, joyful and free labour".[12] One of the speakers even named it the "home sanatorium", as in his view being outdoors in the garden was important in the battle against tuberculosis.[13] The garden should be the place where the whole family could spend their time: the air would benefit the infants and children, the improvement of the family's nutrition by the use of the garden products would ameliorate their health as well, and the garden work would keep the men away from the bottle. In conclusion, gardening was beneficial to all parties involved and all efforts should be made to increase the number of gardens. The congress wanted to propagate the idea by organizing permanent offices that would take care of organizing exhibitions, contests, press releases, articles in journals and magazines, meetings, lectures, regional libraries and bulletins. Many of these ideas were realized in the years to come.

As the questionnaire had been sent to a large number of French cities and to foreign countries as well, the congress report even provides some data on gardens in the French colonies, in eight European countries, in Turkey and in the USA. Data from the Netherlands came from Bruinwold Riedel, but also from De Clercq who had written a letter on the co-operative organization of gardens. In retrospect, Bruinwold Riedel was quite positive about an initiative in Birmingham where labourers grew vegetables and fruit not only for their own use, but for the market as well. He thought this would be possible around the larger Dutch cities (Amsterdam, Arnhem, Rotterdam, The Hague), as the profits made by marketing these products would improve the economic situation of the people.[14] The idea was "to provide every suitable workman with a workman's garden".[15]

Urban gardens

From 1910 onwards, city people became more and more interested in renting a small piece of land in or close to the city. These gardens were called *volkstuinen*, which can be translated as 'urban gardens' as they are not the same as the older *arbeiderstuinen*, the allotments dating from the end of the nineteenth century. The differences between these two types can be listed as follows (see Table 9.2):

Table 9.2 *Comparison between arbeiderstuinen and volkstuinen.*

Arbeiderstuinen (workmen's gardens)	*Volkstuinen* (urban gardens)
main use is to grow food products	main use is to get away from the negative influences of city life; get fresh air, sunlight
a material goal	a non-material goal
area > 150 m^2	area: 100–150 m^2
relatively low rent	relatively high rent
rent to be paid after harvesting	rent to be paid in advance
well-behaved labourers	skilled labourers, teachers, lower-middle-class citizens
– –	sale of grown products is forbidden
no say in the organization	some participation in committees
in rural areas	in cities
– –	governing bodies assess the appearance of the gardens
growing potatoes and vegetables	growing flowers and vegetables
sheds not allowed	garden sheds or even summerhouses

Source: Zantkuijl (1974), 25.

The initiative for urban gardens came from the *Comité voor Volkstuinen* in Amsterdam in 1909.[16] Several distinguished Amsterdam citizens sat on this Committee. They were people like the lawyer, Professor J. A. van Hamel, and the famous Amsterdam architect, H. P. Berlage, people who took an overall interest in social reform. Subsidized by *Het Nut* they founded the first Amsterdam garden plot named *Tuinwijck* (Garden Quarter) in 1910. During the following years three more plots of land were established (*Hofwijck* 1913; *Nieuw-Amsterdam* 1917; *Buitenzorg* 1918). It was characteristic of these early urban gardens that the members of the Committee were still the

Opening of Tuinwijck *the first urban garden complex in Amsterdam by the president of the* Comité voor Volkstuinen, *Professor J.A. van Hamel (10 July 1910).* Source*: Baar (1997), 116.*

well-to-do bourgeois who wanted to improve the standards of living of the working classes, and who acted in the nineteenth-century charity tradition. The garden-holders themselves still had no say in the management and organization of the gardens. Nevertheless the number of urban gardens grew considerably and by 1914 a total of 2,153 gardens covering an area of 40.5 ha was to be found in nine of the larger cities (Amsterdam, Rotterdam, The Hague, Utrecht, Leeuwarden, Groningen, Leiden, Schiedam, Alkmaar).[17]

After the outbreak of the First World War, the City Council of Amsterdam recognized the advantages of these gardens for the people's supply of vegetables and potatoes. However, it took until 1917 before the Council started an experiment for one year in which people could rent a piece of land. More than 144 hectares was farmed out for 3,264 gardens, but although for 600 more gardens there were no applicants. However, the tenants of these municipal urban gardens were the first to organize themselves in a mutual society, the *Bond van Huurders van Volkstuinen* (Society of Urban Garden Tenants), which obtained legal status in 1918. Their goals were more or less the same as those first presented at the international congress in 1903 and have remained unchanged to the present day: to serve the interests of the gardeners, to put out propaganda and give advice on gardening. The president of the society was Jan Vroegop (1877–1946), a policeman, who held the position from 1919 until his death in 1946. The society organized the communal management, purchase of seeds, garden equipment etc., and also fairs and prize contests. One of the first exhibition in the Amsterdam Concert Building in September 1918 gave a lot of positive publicity to their activities,

View of the café at the urban garden complex Hofwijck *in Amsterdam (1921).* Source*: Baar (1997), 117.*

and the organizers concluded that the gardens should "be established permanently". Among the products they showed were cabbages, potatoes, turnips, beans, fruit, and also tomatoes and preserved vegetables.[18] During the following years in other cities (The Hague, Utrecht, Schiedam, Delft, Zaandam, Gouda) similar societies were founded.

The first step to establish a national organization was taken in January 1926. On 26 December 1928 a meeting took place with some ten representatives who discussed the need for such a national organization. On 9 May 1929, organizations from Amsterdam, The Hague, Schiedam and Utrecht agreed on the founding of a national association, even though Zaandam, Delft and Gouda were absent from that meeting. The *Algemeen Verbond van Volkstuinders Vereenigingen in Nederland* (AVVN) (General Association of Urban Gardens' Organizations in the Netherlands) came into existence. The goals of the new organization were the acquisition of legal protection and permanent sites for urban gardens, improvement of gardening methods, education of members in gardening and the publication of a journal.[19] Legal protection became a permanent issue, since as early as 1919 the Amsterdam city council was reluctant to support the gardens, as they were not needed for "additional food supply" any more.[20] The slogan used in the 1920s was: "Urban gardens need a permanent place in the growing city", as none of the Amsterdam gardens remained in the places where they had started ten years before, when the 'war gardens' served to grow vegetables for additional food supplies. The association stressed the important influence on society, the social purposes, the positive effect on family life, the love of nature and, again, air and sunlight, as benefits from the gardens.[21] The ideal garden should be close to people's houses to give the gardener the opportunity to pass by every morning and evening. The first permanent gardens in Amsterdam were established in 1928 and in the same year some gardens were visited by the royal family, which also gave an important stimulus to the movement.

The Amsterdam Society had had its own journal, *De Volkstuin* (The Urban Garden), since 1917 and this publication became the National Society's journal as well: *De Volkstuin. Officieel Orgaan van den Bond van Huurders van Volkstuinen* (The Urban Garden. Official Publication of the Society of Urban Garden Tenants). It started as a monthly newspaper with four pages in a printing of 2,000. In one of the first issues of *De Volkstuin* the society stated that the garden was more than an opportunity for food supply, they wanted to "regain a piece of the beautiful Nature" for their members who were not able to spend their holidays in the mountains or in summerhouses.[22] In 1952 they changed the name of the journal to *De Amateurtuinder* (The Amateur Gardener) that showed the change in their position after the Second World War. They got support from the *Rijkstuinbouwvoorlichtingsdienst* (Governmental Horticultural Information Service) to create this monthly magazine. In the meantime the number of members grew considerably, from 5,840 in 1940 to 21,000 in 1942, when in that year a total of 148 local societies joined in the national organization.

From utility garden to recreation area

As the food supply returned to normal after the First World War, the necessity to grow vegetables and potatoes diminished. So after 1921 the function of the gardens changed: their recreational function became more important than their utilitarian function. However, this was only temporary. During the 1930s, with the increase of unemployment again, more food products were grown while more gardens were created as part of unemployment relief works. Professional growers even felt the gardeners to be their competitors. During the Second World War this attitude changed.[23] The number of urban gardeners rose with an extra 16,500 'war urban gardeners' who wanted to grow vegetables. In 1941, the government decided that seed and fertilizers for the gardens could only be acquired by members of the AVVN. In the April of the following year registration with the National Agricultural Crisis Organization was required and the Director General of Food Supplies had to give his consent for the use of garden products for people's own food supply.[24] At the same time – as people could go nowhere else – the urban garden again became the recreation place for the whole family, with playgrounds for the children and summerhouses in which to spend the night. The use of the gardens as pleasure gardens was described as the situation of the future. A garden with trees, flowers, plants, fruit trees, houses, and with only a few vegetables was seen as the future remedy against a stressful life.[25]

After the War the urban gardens in the cities of Amsterdam, Rotterdam, The Hague and Groningen, having obtained permanent sites, became the recreation areas for the lower-middle-class citizens. Most of these gardens were equipped with a summerhouse while the gardeners grew more flowers and ornamental shrubs than vegetables and fruit. In cities without permanent sites, there were fewer possibilities to create this type of neat looking, and rather costly equipped gardens.

The main shift away from the utility garden started in the 1950s, a period in which major differences between cities were still observed.[26] One of the reasons was the fact that no real profit could be made from the cultivation of vegetables and potatoes. Even in gardens where the whole area was still planted with crops, the owners did not succeed in making money. Around 1970 gardeners explained that if they were to spend their time on some paid labour instead of gardening, they could buy more vegetables etc., but for them gardening had become a hobby. Potatoes were the least profitable, so they changed to vegetables. They even bought a freezer for their winter supplies. In the 1970s, the main purpose of the urban gardens, especially in Amsterdam and Rotterdam, was recreation (see Table 9.3).

Table 9.3 *Growing vegetables and/or flowers in urban gardens in Rotterdam 1958–1972.*

	1958 (%)	1972 (%)
only vegetables	11.4	0.9
partly vegetables	94.0	51.2
flowers	87.0	97.6

Source: Zantkuijl (1974), 84.

The gardens in the urban garden complex Amstelglorie *are used for recreation and not for growing vegetables. Photograph by the author (2000).*

In modern Dutch society "rest and recreation" became the main aims of the urban gardener. Other objectives were "physical exercise" and, for those with a summerhouse, "the social environment". The utilitarian purpose of the garden became less important. Around 1970 many of the gardeners considered gardening as their hobby, and a quarter of them invested money in building a greenhouse in which they could grow better produce. In those years the image of an urban gardener was that of an elderly man with time on his hands. However, a survey showed that especially in those gardens without a summerhouse a considerable percentage of the gardeners were aged below fifty (see Table 9.4).

Table 9.4 *Age groups of urban gardeners around 1970 (n=12,492).*

Age	Amsterdam and Rotterdam with summer-house (%)	The Hague, Utrecht and Groningen with summer-house (%)	Amsterdam, Rotterdam, Hague, Utrecht and Groningen without summer-house (%)	Arnhem and Breda without summerhouse (%)
≤ 24	4	4	5	2
25–34	– –	– –	5	6
35–49	28	25	24	42
50–64	46	42	36	38
65–75	18	24	24	10
≥ 76	2	3	5	1
Unknown	2	2	1	1

Source: Zantkuijl (1974), 89.

Urban gardeners, as they were pictured by the well-known Dutch writer Maarten 't Hart in his book *De droomkoningin* (The Dream Queen), were in general below fifty as well. The principal character in this book, published in 1980, wanders at night through the area of what can be recognized as urban gardens in Schiedam. He is surprised by the "well kept gardens" and wonders how it is possible that there is so much variety on these small pieces of land. He meets people living in the summerhouses even though it is not officially allowed to spend the night in the gardens. These people are trying to escape from the world on the other side of the fence, "that jungle outside" as they call it. Almost all of them are from the lower and middle social classes. There is a majority of women on their own who have run away from home because of marital problems. One of them even predicts that in the year 2020 the Netherlands will consist of cities and urban gardens only, as the families will live in the cities and all divorced people will be exiled to urban gardens.[27]

Present situation

Today the different garden organizations are united in the AVVN with some 26,000 members and gardens having a total area of 4,500 hectares. The AVVN acts as a trade union for hobby gardeners.[28] The continuation of the gardens is their main concern.[29] Throughout the twentieth century there were times when the gardens were threatened with closure or removal and even now the situation has not improved. As many of the gardens are located near the city centres, there is a tremendous pressure from developers to obtain these grounds for building houses, offices etc.[30] The AVVN supports its members in legal procedures to retain their properties. To increase the value of the gardens two new strategies have been developed over the past few years. One is ecological gardening, as in political decisions the 'natural' environment gets priority in protection. The AVVN public relations officer and the AVVN journal *De Amateurtuinder* give advice on gardening without the use of pesticides and artificial fertilizers and on the use of plants and scrubs that increase the diversity of the natural environment and create space for a variety of animals, birds, insects, etc. Another strategy is to open the gardens to a larger public in order to get support from the neighbourhood when the gardens are threatened with closure. The idea is that the more people benefit from the gardens, not only by its produce, but also by being there – outside in air and sunlight – the less chance there is that politicians will take this recreational facility away.[31]

Conclusion

Originally allotment gardening was organized by charities as a means to provide food for the poor labourer and even today these gardens produce an important quantity of vegetables and fruit. However, the need to produce one's own food has changed. The modern urban gardener could buy the greens a lot cheaper in the supermarket, but he and she – as more and more urban gardeners are female – prefer the products of their own work. They enjoy being outside in touch with nature and at the same time they have the opportunity to produce something good. They can make their own choice whether or not to use pesticides and they can pick and eat the vegetables and fruit as fresh as they like.

Equally important is the social structure of the organization of urban gardens. New generations of gardeners want to have the opportunity to use the gardens in the way they please. However, many of the garden organizations are still dominated by the former generation of middle and lower-class gardeners. The communal management, meetings, orderly duties and checks on each other's doings used to be the cornerstones of the urban garden societies.[32] But new gardeners in the Netherlands are a strange mixture of women, foreigners, young urban executives and health freaks. They all have different objectives, some go back to the utility garden, others want a recreation garden only, but many of them want to combine the two. The multicultural Dutch society is represented in the gardens that have been able to survive a century of major reforms in technology and culture. Will the gardens and their inhabitants be able to resist the expansion of the urban environment in the twenty-first century? It will depend on co-operation and, for the gardeners, it will be a ceaseless fight for the retention of their urban gardens. This will be the main issue for the coming years.

Acknowledgement

Mr. H. Vroklage of the *Algemeen Verbond van Volkstuinverenigingen in Nederland* (AVVN), Kemphaanweg 1, 1358 AA Almere, the Netherlands, kindly informed me on recent developments.

NOTES

1. Mijnhardt & Wichers (1984), 105–106, 359–361, 418.
2. Hagoort (1934), 78.
3. Bruinwold Riedel (1905), 101.
4. Ibid., 15–16.
5. Ibid., 117.
6. *Staatsblad* (28 September, 1900).
7. Bruinwold Riedel (1905), 134.
8. Harmsen (1992), 72–75.
9. Rapport *Het belang van de volkstuinen* (1958), 42–43.
10. Tout ce qui concerne la belle oeuvre d'assistance par le travail de la terre, connue sous le nom de Jardins Ouvriers. Lemire & Rivière (1904), 1.
11. Moyens de Développement et de Propagande.
12. De ramener l'homme à la terre, de le rattacher au sol, de lui donner, au sortir de l'usine et de l'atelier sa part de lumière, d'air pur, de travail heureux et libre.
13. Lemire and Rivière (1904), 219, 221.
14. Bruinwold Riedel (1905), 26–27.
15. À chaque ouvrier qui possède l'aptitude requise un jardin ouvrier. Lemire and Rivière (1904), 322.
16. Baar (1997), 114.
17. Oldewelt (1914), 6.
18. *De Volkstuin* 2 (3) (1918).
19. Zantkuijl (1974), 38.
20. *De Volkstuin* 3 (4) (1919).
21. *De Volkstuin* 10 (11) (1927).
22. *De Volkstuin* 1 (3) (1917).
23. *De Volkstuin* 26 (9) (1943).
24. *De Volkstuin* 24 (12) (1941); *De Volkstuin* 25 (4) (1942).
25. Lantinga (1943).
26. Vermooten (1953), 2.
27. Hart (1980), 20, 46, 53, 86, 88.
28. Rapport *Over volkstuinen ... nu* (1978).
29. Rapport *Volkstuinen* (1984).
30. Hellman (1993), 3.
31. Platform Amateurtuinieren (1997).
32. Barning (1992), 37.

BIBLIOGRAPHY

Baar, Peter-Paul de, 'Oude en nieuwe volkstuinders' [Old and new urban gardeners]. *Ons Amsterdam* 49 (1997), 114–118.

Barning, Ton, 'Volkstuintjes als sociale vernieuwing' [Urban gardens as social innovation]. *Intermediair* 28 (35) (1992), 35–39.

Bruinwold Riedel, J., *Arbeiderstuinen in binnen- en buitenland* [Workmen's gardens at home and abroad] (Utrecht, 1905).

Bruinwold Riedel, J., *Tuingrondverhuring aan kleine luiden* [Renting gardens to homely folk], serie 11 (Amsterdam, 1896).

De Volkstuin, 'Tuinen voor het volk!' [Gardens for the people!]. *De Volkstuin* 1 (3) (1917).

De Volkstuin, 'Tentoonstelling met groote keuring' [Exhibition in combination with an important inspection of the products]. *De volkstuin* 2 (3) (1918).

De Volkstuin 3 (4) (1919).

De Volkstuin 10 (11) (1927).

De Volkstuin 24 (12) (1941).

De Volkstuin 25 (4) (1942).

De Volkstuin 26 (9) (1943).

Hagoort, R., *Het beginsel behouden. Gedenkboek van het Nederlandsch Werklieden-verbond Patrimonium over de jaren 1891–1927* [Maintaining the principle. Memorial volume Patrimonium 1891–1927] (s.l. [Amsterdam], 1934).

Harmsen, Ger, *Natuurbeleving en arbeidersbeweging. De Nederlandse socialistische arbeidersbeweging in haar relatie tot natuur en mileu* [Experiencing nature and the labour movement] (Amsterdam, 1992).

Hart, Maarten 't, *De droomkoningin* [The dream queen] (Amsterdam, 1980).

Hellman, Paul, 'Volkstuinders op de schopstoel' [Urban gardeners expelled]. *NRC Handelsblad* (26 June, 1993).

Lantinga, S. S., 'Ontspanningsvolkstuinen' [Urban gardens for recreation]. *De Volkstuin* 16 (4) (1943).

Lemire, l'abbé & Rivière, Louis (eds), *Congrès International des Jardins Ouvriers (Paris, 24 et 25 Octobre 1903)* (Paris, 1904).

Mijnhardt, W. W. & Wichers, A. J. (eds), *Om het algemeen volksgeluk. Twee eeuwen particulier initiatief 1784–1984. Gedenkboek ter gelegenheid van het tweehonderd-jarig bestaan van de Maatschappij tot Nut van 't Algemeen* [For public happiness. Two centuries of private initiative 1784–1984. Commemorative volume on the occasion of the second centenary of the Society for Public Welfare] (Edam, 1984).

Oldewelt, F., *Rapport over den stand der volkstuinen in Nederland* [Report on the situation of urban gardens in the Netherlands] (Amsterdam, 1914).

Platform Amateurtuinieren, *Samen leven Samen beleven. Ideeënbundel voor recreatief medegebruik van tuinenparken* [Living together experience together] (Almere, 1997).

Rapport van de commissie volkstuinwezen, *Het belang van de volkstuinen* [The importance of urban gardens] ('s-Gravenhage, 1958).

Rapport Stichting Recreatie in samenwerking met het Algemeen Verbond van Volkstuindersvereenigingen in Nederland, *Over volkstuinen ... nu* [About urban gardens ... now] (s.l., 1978).

Rapport Algemeen Verbond van Volkstuinders Verenigingen in Nederland, *Volkstuinen Spitwerk in de toekomst. Toekomst-gerichte studie volkstuinwezen* [Urban gardens Digging into the future] (Amsterdam, 1984).

Staatsblad (28 September, 1900).

Zantkuijl, F., *Van coelghaerde tot vrijetijdstuin. Een empirisch-sociologische studie over het volkstuinwezen* [From vegetable garden to recreation garden. An empirical sociological study on urban gardening] (Amsterdam, 1974).

Vermooten, W. H., 'Sociale aspecten van het volkstuinwezen' [Social aspects of urban gardening]. *Socialisme en Democratie* 7/8 (1953), 2–4.

P. J. ATKINS

10 Is it urban? The relationship between food production and urban space in Britain, 1800–1950[1]

The post-modern consumer has become used to the notion of eating 'the other', and exotic fruits, vegetables and other perishable foods from all over the world are now taken-for-granted items on supermarket shelves in west European countries.[2] However, the 'cool chain' and the rapid transport that make the delivery of these items possible over long distances are relatively recent phenomena. In this chapter I want to reflect on an earlier period of history when distance to market was still highly significant, especially for dairy and horticultural produce.

Such was the pull of the urban market and so great the cost and technical difficulty of transport in the nineteenth and early twentieth centuries, that I have found for London that much food production still took place in and immediately around the built-up area.[3] The same was true of other British cities. Commercial cowsheds, piggeries, and market gardens were all common, along with private self-provisioning from allotments and backyard chicken coops and rabbit hutches. The smell of the city must in certain neighbourhoods have had a distinctly rural *bouquet*, from horse manure on the streets, and also from the many slaughterhouses and cattle markets. In short, here indeed was *rus in urbe,* urban farming on a scale that was significant in terms of food output and the visual landscape.

How is it then that at the turn of the millennium we have all but eliminated food production from cities? This transformation has gone so far that urban places are often defined in terms of having 'non-agricultural' land-uses, functions and employment. Also, in rich countries the perception of 'urban' has changed fundamentally in the last 150 years.[4] It may still include an allowance for nature in the tamed and manicured form of public parks and individual back gardens, but noisome and intensive food production somehow no longer seems to be appropriate.

The inspiration for this chapter is two-fold. First, I have spent some time researching food systems in Low Income Countries and I have been struck how different the mental construct of 'urban' is in India and Bangladesh. Cities such as Mumbai (Bombay), Chennai (Madras) or Dhaka owe much of their rapid recent growth to a flood of migrants from rural backgrounds. These people bring with them rural skills that are a useful source of income in their new urban environment, and their desperate need to create a livelihood injects a powerful energy into the lower circuit of the economy that is one factor in the continuance of gardening or milk production in city centres throughout the sub-continent. Faced with such a strong survival of rural functions, one is reminded here of parallels with Victorian Britain, where an extraordinarily rapid and comprehensive urbanization was fuelled by migration on a scale never before experienced, and where urban farming thrived.

Second, there has been a recent revival of interest in urban farming in contemporary social science. Africanists in particular have produced a series of relevant publications. Donald Freeman and Beacon Mbiba, for instance, found spontaneous and vibrant urban farming communities in Kenya and Zimbabwe, and there are many similar examples from around the world.[5] Ignacy Sachs and Dana Silk[6] argue that gardening in tropical cities could produce sufficient food to make a real difference to the nutrition needs for poor people, and there are estimates that 15 per cent of the world's food supply in 1993 came from city plots, rising to a possible 33 per cent in 2005.[7] Globally, 800 million urban dwellers are said to be food producers, mostly for own-consumption but 100 million sell their surplus produce for cash. In some cities the majority of families are involved, from 68 per cent in Dar-es-Salaam to 65 per cent in Moscow.[8] One common finding of this literature is that farming and gardening activities are rarely encouraged by the city authorities, and they are often pushed to the margin of legality in terms of land-ownership and land-use zoning. A second important finding is that food production is a survival strategy for many poor people and it has therefore proved difficult for bureaucrats to eliminate.

Lack of interest until recently in urban agriculture has been due to an absence of official recognition. This has now been rectified, in the international sphere at least, by the Food and Agriculture Organization of the United Nations through three Programmes on: Peri-Urban Production Systems on Animal Production and Health and Veterinary Public Health; Food Supply and Distribution to Cities; and Peri-Urban Horticulture. The international Non-Governmental Organisations have also demonstrated their commitment through the Support Group on Urban Agriculture founded in 1992, the Urban Agriculture Network (1993), and the Global Initiative on Urban Agriculture (1996). Civil society has also generated enthusiasm in this area, such as Canada's 'City Farmer' (started by the government's Office of Urban Agriculture in 1978), which publishes *Urban Agriculture Notes* on the World Wide Web,[9] and also the Urban Agriculture Research Network (AGUILA).

These programmes, networks and databases are intended for the food-deficit countries, but there are also stirrings in the richer part of the world. There is a well-established European network of city farms, which are demonstration projects for educational purposes, and the fashionable theme of 'sustainable cities' has also generated interest in the food-producing potential of wasteland and allotments. Community gardens, as the latter are generically known, are popular throughout Europe, with 30,000 in London and 80,000 in Berlin alone.[10]

The present chapter has two aims. The first is to demonstrate the significance of urban and peri-urban agriculture historically in Britain. By so doing I hope to add to the growing volume of material that argues that cities are not inevitably exclusion zones for farming activity. The second aim is to investigate the specific and contingent conditions under which one activity, milk production, was forced beyond municipal boundaries in selected locations.

Food production in urban and peri-urban Britain

Horticulture

Horticulture was one of the most common peri-urban agricultural activities in and around large British cities. In the nineteenth century there were perishability problems concerning certain fruits and vegetables and market gardens were therefore constrained in their location. Taking London as an example, activity was concentrated within a 25–30 km radius of the city, with some preference for proximity to the Thames, for ease of transport and because of the fertile and well-drained soils of its river terraces.

London's peri-urban horticulture in the nineteenth and early twentieth centuries was highly intensive. Spade cultivation was the norm, made possible by an abundant supply of casual urban labour, but there was much skill to back up the muscle. Many early season and exotic crops (celery, asparagus, melons, pineapples) were forced in the artificial environment of hot-beds or greenhouses and this required expensive and complex systems of support. Inputs of raw materials were also on a large scale, for instance the manure that was essential for enhancing fertility and modifying the heat- and moisture-retention qualities of the soil. Some gardeners are reported to have used 40 tonnes per hectare, and fortunately this was available locally in the form of horse and cow dung from the city itself. One estimate suggests that 250,000 tonnes of cattle and 200,000 tonnes of horse manure were produced in London annually in the 1850s.[11] Much of this was recycled through horticulture.

Table 10.1 *Census enumeration of those employed in horticulture.*

	London	Extra-metropolitan		England & Wales
		Surrey	Middlesex	
1851	9,129	323	3,085	74,324
1871	10,122	5,883	4,573	103,695
1891	11,625	11,931	9,492	179,336

Source: British Population Census.

In 1870 there were said to be 35,000 people employed in horticulture within a 25 km radius of London which, if true, would have represented one-third of the gardeners in England and Wales. Table 10.1 hints that this may have been an overestimate but there were certainly 20,000 in the metropolis of London and the extra-metropolitan parts of Surrey and Middlesex. These people cultivated 5,000 hectares of vegetables, 2,000 hectares of tree and bush fruit, and 400 hectares of herbs[12] in a region dubbed by one writer the "charmed circle".[13]

Milk Production

Peri-urban Britain was also home to the intensive production of milk. I have analysed this in detail for London but we also know that there were cowsheds in other cities.[14] From data published in the *Annual Reports of Medical Officers of Health* we can see that numbers remained substantial well into the 1920s.

Administrative boundaries present us with a problem in gauging the scale of production. Although urban districts were often given substantial stretches of open countryside when first established (overbounded), later the process of urbanization often saw the built-up area extend beyond the city limits (underbounded). Boundaries were adjusted to take account of this expansion but we should be wary nevertheless of the data in Table 10.2 which purports to compare the milk-producing capability of cities through the surrogate of the cows within their jurisdiction.

Table10.2 *The milk-producing capacity of selected cities, 1911–1913.*

Town/city	Area (ha)	Cows	Human population
Birmingham	13,487	575	525,960
Bradford	22,881	4,400	288,505
Burnley	4,005	726	106,337
Halifax	13,983	2,500	101,556
Leeds	21,593	2,000	445,568
Liverpool	16,642	6,428	746,566
London	74,816	3,096	4,522,961
Manchester	21,645	1,943	714,427
Newcastle	8,452	497	266,671
Norwich	7,896	567	121,493
Nottingham	10,935	860	259,942
Preston	3,971	304	117,113
Salford	5,202	195	231,380
Sheffield	23,662	2,400	454,653
Sunderland	3,357	171	151,162

Source: Medical Officer of Health Annual Reports of the various cities.

As with horticulture, the best definition of 'urban' milk production is the degree of its intensification. In some cities this was taken to the extreme of stall-feeding, with many cows being kept indoors for the full length of their lactation and then disposed of to the butcher. The maximum possible yield from each cowshed was extracted by cramming the animals together and feeding them a stimulating diet that included spent brewers' grains. This approach was most common in London, Liverpool, Glasgow and other cities where milk had long distances to come from rural suppliers. Table 10.3 suggests that around the time of the First World War there were still cities that continued to source milk from within their own boundaries.

Table 10.3 *The proportion of milk sourced from cows kept within the urban boundaries.*

Town/city	Date	Urban production (% of total supply)
Belfast	1929	20.0
Birkenhead	1919	20.6
Bootle	1918	41.0
Bradford	1920	61.5
Edinburgh	1921	52.0
Folkestone	1915	25.0
Liverpool	1927	29.5
London	1850	80.0
London	1880	28.3
London	1910	2.80
Newcastle	1925	15.0
Sheffield	1929	24.8
Weymouth	1914	33.0

Source: Mainly from the Medical Officer of Health Annual Reports of the various cities.

Pigs

Small livestock (pigs, chickens, rabbits, pigeons, bees) were also present, mainly as a backyard supplementary food source for working people, but we have no data on this, other than a few scraps of information. The most tolerant city seems to have been Liverpool, where 5,000 pigs were kept in licensed commercial pigstyes in the 1930s. Harry Thorpe's survey of 1969 showed that animals were still frequently kept on allotments in towns but he found that this caused friction about smells, flies, dirt and "disreputable structures" that "lower the tone" of sites:

> The tenants of allotment gardens whose only desire is to grow fruit and vegetables on their patches of ground, assert that the allotment holders who keep livestock are different people; they have different aims, different interests and different standards.[15]

Allotments

Until about 1900 allotments in Britain were mainly rural, providing farm labourers with sustenance for their families. The Small Holdings and Allotments Act (1908) seems to have changed this, forming the basis of modern law in this area and giving encouragement to the provision of urban plots. The Allotments Act (1925) required allotments to be considered in every new town planning scheme. The First and Second World Wars, with their campaigns such as "Dig for Victory", were further stimuli. In 1939 there were 570,000 plots in the urban areas of England and Wales, and in 1944 these were estimated to be providing 10 per cent of national food needs.[16]

The decline of urban food production

Bid-rent theory would have us believe that agriculture inevitably falls before the onslaught of bricks and mortar, because the latter will always be more competitive in bidding for land. Although this is essentially true, the fuzziness of the land market in the real world prevents the fully rational outcome anticipated by such normative economic models. It seems that many other factors have been involved in the decline of urban farming.

As far as horticulture is concerned, for instance, tenurial conditions were always important in the peri-urban zone. Leases were often short and market gardeners had to calculate whether it was worth their while to occupy a particular piece of land for the time that was available. Thus tree crops were never popular close to the suburban fringe for the obvious reason that the 10–15 year lag before fruiting might be not be completed before the plot was ripe for housing or industry. Landowners often included a "resumption clause" in leases, which meant that possession could be resumed "at will", usually when a suitable offer came in from a builder.[17] Vegetables were ideal for short leases where construction was anticipated because yields were high and catch crops could be taken in weeks. Growing them therefore helped avoid the need for land to stand idle between land uses, which is a common feature of the so-called urban shadow.

Intensive production declined near London around 1900 for a number of reasons. First, the cost of casual labour rose, making weeding and harvesting more expensive and at the same time one the main sources of cheap fertility was choked off as horses were replaced by the internal combustion engine and cows were banished from the city. Second, there were problems of smoke pollution and trespass that made inner-city locations increasingly difficult for market gardeners. Third, there was increasing competition from producers within a few hours carting range who farmed at a lower level of intensity and therefore lower cost, and from provincial areas with some natural advantage or historical specialism. Kent, the Vale of Evesham, the Fens, Bedfordshire and coastal Cornwall were all-important examples of the latter. The advent of the railways enabled their competitive edge to sharpen in the last thirty years or so of the nineteenth century.[18] Imports of foreign fruit also increased.

The idea of urban horticulture has not faded entirely. Many cities still have nurseries within their boundaries and all have garden centres. The emphasis has now shifted from food production to ornamental plants and much of the stock comes from out-of-town wholesalers.

The idea of animal husbandry in cities has been more thoroughly challenged. Chris Philo, in a welcome excursion into the relationship between animals and the 'urban', has argued that among the inclusions and exclusions, non-pets have been assigned by the modern mind to the latter category because of their perceived transgressions against the norms of cleanliness and purity.[19] The lives of food-producing animals are now seen to belong to the 'rural' world, and even their deaths in urban slaughterhouses are less tolerated. The construction of this attitude is worth discussion because it was complex and not without resistance.

The sanitary discourse of Victorian and Edwardian Britain was a powerful vehicle for eliminating from towns those activities associated with smells and dirt, and connected in many people's minds with disease. In the capital the process was already under way in 1853 when the new Medical Officer of the City of London, John Simon, introduced draconian bye-laws about the construction of cowsheds. It was the wealthy residential districts of the West End, however, which were most assiduous, with several local authorities in the 1860s and 1870s refusing to licence cow-keeping within their jurisdictions. No doubt local lobbying was a factor, although we have no evidence of this. Withholding licences was possible within the Provisions of the Metropolitan Self Management Amendment Act of 1862 but most of London was not greatly affected until 1879 when the Metropolitan Board of Works made regulations under the Dairies, Cowsheds, and Milkshops Orders. From then on pressure was applied by requiring improvements in such aspects of cowshed construction and management as paving, lighting, clean water provision, the volume of air space per animal, ventilation, drainage and cleansing. The additional expense of these alterations was sufficient to force many milk producers out of business, especially as country milk brought by railway was becoming more competitive in the last few decades of the century.

From this account one might imagine that the creation of a sanitised city, devoid of food-producing animals and noxious trades such as slaughtering, tanning and soap boiling, would only have been a matter of time. But the reality was complex, with variations both in time and space.

In London, the East End authorities seem to have been the most tolerant of cows and pigs. Food production was a significant activity in districts such as Whitechapel and Bethnal Green, both poor areas where sustainable livelihoods could not be taken for granted. Closing off such an opportunity for employment would not have been appreciated by voters. Also, here the inhabitants of the slums were much less vocal about environmental hazards than their more comfortable counterparts in Belgravia and Mayfair, and the low rateable value of property meant that the Sanitary Authorities had less income to invest in inspectors and prosecutions.

In other parts of the country there were also variations of practice. In most cities cowkeepers continued longer than in London. In Liverpool, for instance, there were still 1,000 urban cows as late as 1956 (see Table 10.4),

Table 10.4 *Cows and cowsheds in Liverpool.*

	Licensed cows	Licensed cowkeepers
1870	4,948	493
1880	5,322	450
1890	5,104	378
1900	5,905	437
1910	6,691	447
1920	4,942	295
1930	4,931	281
1940	3,644	187
1950	1,882	99
1960	519	23

Source: Medical Officer of Health Annual Reports.

and this was undoubtedly because of the Corporation's conviction that cattle under their control were less likely to be diseased than those based rurally. Table 10.5 shows that they were correct.

Table 10.5 *Tuberculosis detected in Liverpool's milk supply, 1896–1950.*

Urban milk		Country milk	
Samples	% tuberculous	Samples	% tuberculous
20,139	4.30	24,330	6.39

Source: Medical Officer of Health Annual Reports.

City authorities were understandably wary of the quality of the rural milk sent to their citizens. In the late nineteenth century this was very likely to be adulterated, as well as dirty and possibly diseased.[20] The housing conditions of cattle were appalling (see Table 10.6) and there was little enthusiasm for imposing expensive improvements:

> The urban [and rural] district councils are dominated by uninstructed farmers who regard regulations governing milk supplies as nuisances to themselves. The inspectors are discouraged from performing the work adequately, and if there is an inspector in those areas who does his work conscientiously he will not long keep his job.[21]

Table 10.6 *The state of country shippons serving Manchester.*

	No cows with tuberculous udders	Cows with tuberculous udders
Bad general condition of building	37	56
Size and ventilation insufficient	34	62
Shippon not clean	28	53
High proportion of old cows	27	53
Cows in poor condition	13	21
Average number of cows in suspicious udders	5.6	12.0
Average number of cows with tuberculous udders	0.0	3.7
Average number of cows per farm	20.0	25.0

Source: Delépine (1910, 1332).

Part of the problem was that before the 1920s rural producers were not penalised for poor quality milk and were not under any pressure from their local authorities. It is scarcely surprising therefore that urban cowkeepers were finding it difficult to compete. The situation was summed up in the House of Commons by Dr Christopher Addison:

> The members of the various authorities in the country, that is, small urban councils and so forth, frankly, are not enthusiastic to adopt the Cowsheds Orders; are not enthusiastic to secure clean milk supplies for London or Birmingham, or some place two or three hundred miles away. When one comes to study the various Public Health Acts and the enormous number of Private Acts which have to do with the control and regulation of our milk production, I think it is fair to describe them as a hopeless chaos.[22]

The rural regulatory laxity demonstrated in Table 10.7 was changing in the 1920s and 1930s but by then it was too late for many city cowkeepers. Those who survived were largely to be found in towns where fear about tuberculosis in country milk outweighed the urge to cleanse the urban built environment.

Table 10.7 *Cowshed regulations adopted by 1907.*

	Regulations under Article 13 of 1885 Order	Regulations based on Local Government Board Circular of 1899	No regulations
County Boroughs	39	30	4
Other Boroughs	86	127	40
Urban Districts	330	345	131
Rural Districts	220	266	171

Source: Ministry of Health (1906–1909); Return showing names… (1907).

Conclusion

In the twentieth century urban food production declined steadily in Britain. There was competition from rural and foreign producers, both of whom had lower overheads and whose comparative advantages were facilitated by improved transport and storage technologies and also by better organized marketing channels. Meanwhile the costs of urban producers rose, not least because of the regulatory framework established by the local state. However, that local state was not monolithic by any means. The Dairies, Cowsheds and Milkshops Orders (1879, 1885, 1886, 1889, 1899) were permissive and inevitably this led to a great deal of variation between urban areas, let alone between town and country. It was not until the Milk and Dairies Order (1926) that inspections became compulsory and conditions of production became subject to general standards.

Market gardening, cow and pig keeping were increasingly marginalized in urban life. It is noticeable that animals which had no access to grazing were concentrated in the poorer districts, where complaints about smells and other forms of pollution were rare. Their keepers were often poor people, especially rural immigrants finding their way in a strange environment. Horticulture also survived in the peri-urban zone longest in the interstices of the built fabric. This was not resistance to the flood tide of urbanization from a group of eccentric gardeners but rather a means of optimizing returns to land that everyone knew would eventually be developed. There was an economic logic here which suited the growers, the landowners, and presumably also the consumers but, as the opportunities faded, so did the production.

The power of the urban land market and of the officious state have swept British cities clean of farming. The survival of allotments, garden centres and pony clubs is a modest reminder of the country in the city but the memory is dim because the urban mind itself has effectively eliminated the very idea of food production in its area of dominance. Urban farming has been thought out of existence and banished to an outer sphere, the mutually exclusive binary category that we call 'rural'. This outcome is not inevitable. As suggested in the introduction, urban farming continues to thrive in Low Income Countries and there are currently international efforts to encourage it further. The hygienist discourse, which has dominated much Western thinking about urban planning, is less evident in the poor world, and urbanization there is still a magnet for people with rural skills. One imagines that it will be several decades yet before the milk producers disappear from Madras or Dhaka.

NOTES

1. I am grateful to the participants at the Sixth Symposium of ICREFH in Tampere for their comments on a earlier draft.
2. Atkins & Bowler (2000).
3. Atkins (1977); Atkins (1987).
4. The classic reference for this is Williams (1973).
5. Freeman (1991); Mbiba (1995).
6. Sachs & Silk (1990).
7. FAO; Faruqui et al. (1998).
8. Helmore & Ratta (1995).
9. Faruqui et al. (1998), http://www.idrc.ca/cfp/rep22_e.html.
10. Meikle (1999).
11. Atkins (1987).
12. Cuthill (1870); Shaw (1879).
13. Whitehead (1878). Stanhill (1977) gives a fascinating account of the equivalent *marais* in Paris that shows how extraordinarily intensive market gardening was in the nineteenth century.
14. Atkins (1977).
15. *Report, Departmental Committee of Inquiry into Allotments* (1969), 192.
16. Ibid.
17. Bennett (1950).
18. Atkins (1985).
19. Philo (1995).
20. Atkins (1991); Atkins (1992).
21. *Parliamentary Debates* 161 (1923), 194–195.
22. *Parliamentary Debates* 39 (1912), 190–195.

BIBLIOGRAPHY

Atkins, P. J., 'The intra-urban milk supply of London, circa 1790–1914'. *Transactions of the Institute of British Geographers,* new series, 2 (1977), 383–399.

Atkins, P. J., 'The production and marketing of fruit and vegetables 1850–1950', in: D. J. Oddy & D. Miller (eds), *Diet and Health in Modern Britain* (London, 1985), 102–133.

Atkins, P. J., 'The charmed circle: von Thünen and agriculture around nineteenth century London'. *Geography* 72 (1987), 129–139.

Atkins, P. J., 'Sophistication detected: or, the adulteration of the milk supply, 1850–1914'. *Social History* 16 (1991), 317–339.

Atkins, P. J., 'White poison: the health consequences of milk consumption'. *Social History of Medicine* 5 (1992), 207–227.

Atkins, P. J. & Bowler, I. R., *Food in society: economy, culture, geography* (London, 2001).

Bennett, L. G., *The development and present structure of the horticultural industry of Middlesex and the London region* (unpublished University of Reading Ph.D., 1950).

Cuthill, J., *Market gardening, or the various methods adapted by gardeners in growing for the London market*, 5th edn (London, 1870).

Delépine, S., Report to the Local Government Board, on investigations in the Public Health Laboratory of the University of Manchester upon the prevalence and sources of tubercle bacilli in cow's milk, *Report of the Medical Officer to the Local Government Board, 1908–1909*, Appendix B, no. 5 (1910), 341–414.

FAO, Issues in urban agriculture, *Agriculture* 21. http//faoext02.fao.org/ WAICENT/ FAOINFO/AGRICULT/magazine/9901sp2.htm.

Faruqui, N., Mougeot, L., Smith, O., Wilson, B. & Hovorka, A., Cities feeding people: programme summary, 1997–2000, CFP Report Series, Report 22 (Ottawa, 1998). http://www.idrc.ca/cfp/rep22_e.html.

Freeman, D. B., *A city of farmers: informal urban agriculture in the open spaces of Nairobi, Kenya* (Montreal, 1991).

Helmore, K. & Ratta, A., 'The surprising yields of urban agriculture'. *Choices* (UNDP) (April, 1995), 22–27.

Mbiba, B., *Urban agriculture in Zimbabwe: implications for urban management and poverty* (Aldershot, 1995).

Meikle, J., 'Pigs in the city'. *Guardian* (11 August, 1999).

Ministry of Health (1906–1909), *Milk Bill Papers 1906 to 1909*, Public Record Office MH 80/3.

Philo, C., 'Animals, geography, and the city: notes on inclusions and exclusions'. *Environment and Planning D: Society and Space* 13 (1995), 655–681.

Report, Departmental Committee of Inquiry into Allotments (Cmnd 4166), Chairman: H. Thorpe (London, 1969).

Return Showing the Names of the Councils of Boroughs and Urban and Rural Districts who have Made Regulations under the Dairies, Cowsheds, and Milkshops Order of 1885, *British Parliamentary Papers* 1907 (152) lxvii. 381.

Sachs, I. & Silk, D., *Food and energy: strategies for sustainable development* (Tokyo, 1990).

Shaw, C. W., *The London market gardens: or, flowers, fruits, and vegetables as grown for market* (London, 1879).

Stanhill, G., 'An urban agro-ecosystem: the example of nineteenth-century Paris'. *Agro-Ecosystems* 3 (1977), 269–284.

Whitehead, C., 'The cultivation of hops, fruit and vegetables'. *Journal of the Royal Agricultural Society of England*, series 2, 14 (1878), 723–760.

Williams, R., *The country and the city* (London, 1973).

AURA KORPPI-TOMMOLA

11 Hungry, but not starving

The experiences of Finnish children in town and country during the Second World War

During the Second World War Finland was twice in arms against Russia. During the first conflict in the Winter War (1939–1940) Russia conquered the eastern parts of the country, the provinces known as Karelia. In the War of Continuation (1941–1944) Finland reconquered Karelia, but lost the area again in the Truce of Moscow in September 1944. Even after the truce agreement was made, the Finnish Army still had to drive the Germans out of Northern Finland during the winter of 1944–1945. The civilian inhabitants of Karelia were evacuated many times until they were finally settled as displaced persons in the smaller but still independent Finland. About 88,000 men were killed in action and after the war there were 30,000 widows and 50,000 war orphans in the country.

A total of 70,000 Finnish children were evacuated to Sweden and Denmark during the war in order for them to obtain better medical care and nutrition. Most of the historical research regarding children in wartime has been concerned with the experiences of these evacuees.[1] What happened to the majority of Finnish children, did not seem to interest anyone until in a seminar in 1995 a woman, who herself had been sent to Sweden during the war, asked: "But how was it for the rest of the children in Finland? Was it any better to stay at home and be afraid than to be sent to Sweden and miss one's parents?" Her questions aroused our interest in the war experiences of Finnish children. Another motive for undertaking this research was the publication of the book by Elina Haavio-Mannila and Riikka Raitis on women during the war years. In the course of their study they found that it was no longer possible to contact some groups of women, for instance, the mothers of fallen soldiers.[2] We decided to start collecting memories, while wartime children are still among us. The same pattern (to focus on a minority of children) seems to have occurred in other countries as well as Finland. In the United Kingdom some research has been done on children who were evacuated to other parts of Britain or even other parts of the Empire during the war.[3] Charles Perkins has made a comparative study of children in wartime by collecting childhood memories from every continent.[4]

We decided to carry out a national survey to inquire into the living conditions of children during the war years.[5] Two history students, Pekka Haavisto and Taru Järvenpää, undertook this survey seeking answers to such questions as: were children hungry or undernourished? What did they eat? What were everyday meals like and what was served on special occasions or at parties?[6] A pilot study was undertaken first from a small sample of people in the target group.[7] These people were from two provinces: Satakunta in the west and Kymi in the eastern part of Finland. Contact was established with them via the Mannerheim League for Child Welfare, which is a nation-wide non-governmental organization. Using the pilot study, questions for the main national survey were formulated, and a questionnaire was sent to 800 Finns born between 1927 and 1938.[8] We received 273 replies to the main survey, i.e. 34 per cent of the forms that were sent out. Responses came from all over the country. In 260 returns there were comments in reply to the questions concerning food. Relatively few men, only 36 per cent, had answered the pilot study questionnaire, but in the final national inquiry 45 per cent of the respondents were men.[9]

From this research it is impossible to generalize or draw exact conclusions about childhood experiences during the war years. We merely have fragmentary information from different regions of the country. Of the 800,000 Finns born in the 1927–1938 cohort more than 200,000 had already died before we sent out our questionnaire. Many of the remaining people were too ill to be able to answer the questions while others were unwilling to do so. Some of them replied that they did not want to be reminded of those unhappy years. The questionnaire also took considerable time to answer and this may have been part of the reason for the low number of returns. However, the sample size did not concern us over much, because it was our aim to collect some memories from the Finns who lived through the war years as children. The material is valuable in its own right, as Natalie Zemon Davis points out in her article *Who Owns History*? She wants to widen the scope of history by expanding our knowledge of those groups silenced by some form of oppression, as in the case of particular populations or religious groups.[10] In our research it might be said that children's experiences have been subordinated to adult memories.

Not starving but hungry

During the war, there was enough to eat as a rule in most families (211 of 260 answers). There was no significant difference between men and women. Just under 40 per cent of respondents recalled that daily meals consisted mostly of bread, porridge and gruel. Meat and fish were mentioned only rarely: both in six per cent of the responses. Less than 20 per cent of the respondents remembered or mentioned that their family supplemented their diet with berries and mushrooms. Nevertheless, just over 30 per cent described their everyday food as "normal", which must mean that it was not different from the meals they had before or after the war. But there were times when children were hungry. Most of them got enough food to survive, but only four per cent of our respondents do not remember suffering from hunger; 45 per cent replied that they were always or very often hungry. It was especially difficult for young boys (see Table 11.1). They had to work like men for which they needed energy but they also needed nutrients for growth. A male respondent recalled: "there were days when we had nothing to eat" (m5/1938). [11] A woman remembers: "a growing child would have eaten more, but we had strict meal times and snacks were not allowed in between"(f12/1935). Other comments highlighted the shortages of food:

> When the flour was running out we boiled green peas and baked them with the last flour. It did not taste good. (m178/1936)

> When we had used all of the food rationing coupons, we often had it difficult. Sometimes we ate plain potatoes. I remember that once we boiled pea pods and ate them to satisfy our hunger. Growing children were always hungry. Mother many times gave her share to us children. (f45/1932)

One woman remembers being hungry during the evacuation from the war zone in 1939. Her family travelled in a horse-drawn carriage through southern Finland and did not always find a place to stay for the night (f16/1933). The most touching stories are from the cities where it was neither easy to obtain extra food nor money to buy even the rations to which they were entitled. All kinds of unofficial bartering and exchange took place and the children learned it, too.

Table 11.1 *When hungry?*

	Number	(%)
Always	18	23
Very often	17	22
Sometimes	30	38
Other	10	13
Never	3	4
Total	**78**	**100**

Source: Questionnaire.

There were two significant factors affecting the child; children in the cities were hungrier than children in the countryside. The social status of the parents was also a significant determinant which separated the families in many respects. Hungry children had other worries, too. Taru Järvenpää found that social background was the key explanation differentiating children's success in education.[12]

Food shortage was especially pronounced in the urban areas and children living in them suffered accordingly. It meant meagre meals and hunger particularly for young people but it was far from starvation. One gets the impression that in many answers people are describing the same kind of food shortage, only individual approaches vary – some have more bitter memories about it than others. The supply situation in Finland for foodstuffs was most difficult during the last winter of the war, 1944–1945. However, this was not apparent from the study. In response to the question when they felt hungry answers gave the time of day rather than the season. They simply tell us that children were hungry "often in the night". A girl might have felt hunger when she had to go to bed. She missed her father and could not sleep (f1927/32). One man remembered that in the city of Helsinki food was "extremely simple, for instance, it was soup boiled out of the leaves of Swedish turnips and milk or salted mushrooms with potatoes, occasionally fish. The lake was three kilometres away, so the produce from its waters were not available daily"(m173/1929)

Life was not rosy out in the countryside either: "In the early summer food was bad, when potatoes had frozen the summer before and the little we had had been used up/gone rotten, and the cows were dry. We had to walk around looking for possibilities to buy skim milk – a pail for milk in one hand, some money in the other and our hearts in our mouths."(m252 /1927) Neverthless, in the countryside in western Finland far away from the front a farmer's son describes there being plentiful food. In a farmhouse "for breakfast we ate potatoes, fried pork gravy, or pork in brine maybe gruel or oatmeal porridge. We had even a better lunch with butter, goat's milk cheese, boiled or fried eggs, milk and sometimes home-brewed beer or buttermilk on the table." In the province of Satakunta the lunch in the summer consisted mainly of gruel and porridge from a powdered mixture of dried oats, barley and peas (*talkkuna*). When the daily work was done dinner was heavy and consisted partly of leftovers from the day. Coffee or coffee substitute was served at 10 a.m. and 4 p.m., in the morning with sugarless 'sweet' buns (*pulla*) and with bread without butter in the afternoon. "Rye or wheat bread was always on the table." (m148/1933) However, quite close to him lived another boy, who had "hunger often visiting his body, but there was enough of basic food, bread, potatoes and quite often even meat or fish. We gave food to people poorer than we were and to soldiers, too. At least once my father sent elk meat from the front in Eastern Karelia. It must have been the oldest inhabitant of the forest, because it took four hours to cook."(m53/1930)

In the countryside there was often even a surplus of food to give to family members living in cities. Children were sent to visit family members in the countryside: "My uncle had a farm and we visited him in the summer. There we could get our stomachs full. To compensate for it we helped to rake up hay, whisked off flies from the cows when they were milked and weeded the garden. We picked some berries, too. We got food to take back home." (f45/1932) Many Finns remember that their parents made trips to the countryside to buy food from the farms, often from farmers to whom they were related or whom they had known already before the war (see Table 11.2). There were 88 respondents who remembered or knew of such trips against 107 who replied that they did not remember and 78 who did not answer at all. Of the children born between 1927 and 1938 only 49 remember travelling with their parents to get food from the countryside. They understood that it was illegal, because the farmers were obliged to sell everything to the authorities who would then sell it in strictly calculated food rations.

Table 11.2 *Trade on the black market.*

Response category	Buying at the black market		Trips to buy illegal extra food	
	Number	(%)	Number	(%)
Yes	107	39	88	32
No	109	40	107	39
No comment	57	21	78	29
Total	**273**	**100**	**273**	**100**

Source: as Table 11.1.

The trips were very exiting. Among the responses we find many stories about how people avoided controls in the buses and trains. Young boys walked or skiied tens of kilometres to get their rucksacks full of bread and returned late at night. Someone hid butter in a baby carriage. The black-market goods were paid for with money though often the deal was made by exchanging, say, bread and flour against Wellingtons or the best of all barter 'currencies', cigarettes. Nobody seems to have had any doubts about doing it or a bad conscience about it, so great was the need for extra food. Young children travelled with their mother or elder siblings, but at the age of 10 for boys and 14 for girls, children sometimes made these trips alone.[13] The fear and great efforts described in some war time memoirs cannot be found in these short answers. Some respondents say their family did not buy anything in the black market, because it was a family policy to remain law-abiding citizens. Some of them just comment on the lack of money as the reason for not being tempted. Children in the countryside remember people coming to ask for food. There were farmers who gave their siblings and cousins living in the cities a parcel of land to grow their own potatoes and vegetables.

Consumption of alcohol by grown-ups

The strain of war made many people try to calm their nerves by stimulants (alcohol) or sedatives (cigarettes). So, we asked respondents at what age they tasted alcoholic beverages for the first time (see Table 11.3). We got 240 answers. As many as 26 people (10 per cent) said that they do not use alcohol at all and never did. This is about the same proportion of Finns who do not drink alcohol even today (nine per cent of male population and 13 per cent of female population according to The National Public Health Institution).

In Finland the age of fifteen is a turning point from childhood to adult life, when young people are confirmed and are allowed more freedom to act independently. Only 14, or five per cent, of our wartime children replied that they had their first drink before that age; 24, or nine per cent, say they became acquainted with alcohol at the age of 15. The largest number, or 40 per cent drank before they turned 20 while the remainder first drank before the age of 25 years. Many people commented on the particular occasion when they had their first drink even though we did not pose this question. Very often it was taken at a high school graduation party, wedding or other special occasion, when respondents felt they were allowed to be adults. All in all, alcohol was clearly a part of adult life for that generation, and did not play any role in their young lives as children.

Table 11.3 *Age when alcoholic beverage first consumed.*

Age	Number	(%)
Before 15	14	5
15	24	9
Before 20	109	40
Before 25	67	24
Never	26	10
No comment	33	12
Total	**273**	**100**

Source: as Table 11.1.

Provincial delicacies and coffee substitute at parties

An interesting part of the results is the information regarding the food served at get-togethers and family reunions. Only 170 persons answered the question about parties. Of these, 118 of them, 74 women and 44 men, describe social occasions such as parties, weddings, and funerals. Girls helped in the kitchen more often and this may be the reason why they remember the parties better. Almost one-third of the 170 respondents, 25 women and 27 men, do not remember being at any parties at all, or maybe once "when an uncle sent a gift from the United States" (m53/1930).

The dishes did not seem to vary from weddings and birthday parties to funerals. Some respondents tell us that the served food was the same every day, but sometimes it was just called special: "green pea soup" one day and "funeral soup" the next day for funeral guests. In general it seems that those children who helped in the kitchen remembered more of what was served than others. On the other hand the hungriest ones remembered those times when they got some special treats.

The responses describe the variety of traditional Finnish dishes. What is significant is the way that provincial delicacies occur. On the west coast, ham and salads of pickled herrings and vegetables were common. The central provinces, including Häme, were the areas for casseroles. In the eastern parts, and among the evacuated population, Karelian rice pastries and Karelian mixed casserole of beef, pork and mutton were dishes for parties, but sometimes also in everyday life (see Tables 11.4 and 11.5).[14] The three

Table 11.4 *Food served at parties.*

	Number	(%)
Karelian pastries or casseroles	11	4
Casseroles	8	3
Vegetable salad and ham	12	4
'Sweet' buns (*pulla*)	33	12
Other servings	54	20
No parties at all	52	19
No comment	103	38
Total	**273**	**100**

Source: as Table 11.1.

families on the west coast who served Karelian dishes, were evacuated from the war zone in the east. The several evacuations of people from the eastern parts of the country, Karelia, meant a positive change for children, although it was difficult and painful for adult people. Many children in the western and central parts of the country learned to eat the Karelian rice pastries from evacuated families. After the war years many of the provincial dishes have become known all over the country and they have turned into more 'Finnish delicacies' than local specialities. The arrival of evacuees influenced the life of children also in other aspects: the respondents commented that they were happy to get new playmates from among the evacuated children in their neighbourhood.

Table 11.5 *Food served at parties in different parts of Finland.*

	Karelian pastries/ casserole	Casseroles	Vegetable salad and ham	'Sweet' buns (*pulla*)	Other servings	Total
Eastern provinces	6	2	4	10	13	35
Western & central provinces	3	3	5	8	17	36
Northern & southern provinces	1	--	--	9	15	25
Not specified	1	3	3	6	9	22
Total	**11**	**8**	**12**	**33**	**54**	**118**

Source: as Table 11.1.

Although provincial differences and a variety of Finnish traditional dishes served were shown by the answers, the most typical gathering was the coffee party where substitute coffee made from domestic ingredients and 'sweet' buns (*pulla*) were served. On some occasions, even the 'sweet' buns were made of rye instead of wheat. Guests were asked to bring their own sugar and cream. Sometimes there was not much to offer, but maybe some juice and biscuits made the gathering a party. It looks as if families saved their rations for family celebrations, parties and special events. These occasions were a welcomed change from everyday life, and children had pleasant memories of them. Only the funerals were unsettling, because they were the first time children met adults crying and didn't understand the grief themselves.

Contentment from simple pleasures

At the end of the long questionnaire we let the respondents describe what they thought war meant to them; whether it changed the course of their lives, and whether they were content with their lives. Their message is that they learned to be satisfied with little things and to enjoy simple things like daily meals. There are some dishes that the children learned to hate during the long years of war. It also left them a yearning for sweets and candies, which were unavailable during the war. The rationing of food lasted after the war during the reconstruction period until the middle of the 1950s. So these memories are certainly a mixture of stories of many years of lack of sufficient food.

It is also evident that those respondents who had been content during the war had lived in families who kept together and where adults did not share all their worries with their children. On the contrary, children were left to play and help in the daily tasks at the household. Children who were loved and protected from the cruelties of life tell us their survival stories proudly, for instance: "My mother was an excellent cook and made miracles out of nothing." As Pekka Haavisto found in his part of the research most of the families did not exchange their fears and uncertainties, but wanted to protect children from the evils of the world around them.[15] On the other hand, in families with desperate parents little food seemed even less. Similar findings were made by Ilse Eriksson-Stjernberg in her study of adults' memories of the war they had lived through as children. Emotional insecurity was strong among the people, whose parents or guardians had shown their bitterness to their children or shared it with them.[16] Loving and responsible adults seemed to be the key to a happy childhood during the war and overcame food shortages.

NOTES

1. Evans (1984); Kavén (1983); Korppi-Tommola (1996).
2. Haavio-Mannila & Raitis (1993), 21, 248.
3. Bean & Melville (1989); Jackson (1985); Stokes (1994); Wicks (1988).
4. Perkins (1998).
5. We ascertained that 800,000 children had been born in Finland between 1927 and

1938. This was our target population: when the war began they were aged between one and 12 years old; when it ended in 1945, their ages ranged from seven to 18 years. We set out to obtain a 1:1000 sample from this birth cohort and so, in the spring of 1997, a questionnaire was sent to 800 Finns born between 1927 and 1938.

6. These questions on food were amongst a variety of topics asked.
7. The pilot study was based on 58 responses.
8. We received their addresses from the Population Register Centre.
9. The original material of this research is at the head office of the Mannerheim League for Child Welfare, Helsinki, Finland.
10. Zemon Davis (1999), 20–22.
11. The returned forms are numbered. In the quotations I have indicated the sex – male or female – plus the number of the form/year of birth. (m5/1938) means a man born 1938, form number 5.
12. Järvenpää (1999), 116.
13. Fagerlund (1994).
14. Hietala (1997), 232.
15. Haavisto (1999), 100–101.
16. Eriksson-Stjernberg (1996), 31.

BIBLIOGRAPHY

Bean, Phillippe & Melville, Joy, *Lost Children of the Empire. The Untold Story of British Child Migrants* (London, 1989).

Eriksson-Stjernberg, Ilse, 'Mitä lapsena koetusta sodasta mielessä aikuisena' [What recounts adult persons have of the war they experienced as children], in: Aura Korppi-Tommola (ed.), *Sotalapset* [War children] (Helsinki, 1996), 20–45.

Evans, Alvin L., *Ero ja lapset* [Separation and children] (Helsinki, 1984).

Fagerlund, Bengt, *Sotaa ja leikkisotaa* [War and play-war] (Porvoo, 1994).

Haavio-Mannila, Elina & Raitis, Riikka (eds), *Naisten aseet : suomalaisena naisena talvi- ja jatkosodassa* [Weapons of women] (Porvoo, 1993).

Haavisto, Pekka, *Sota lapsen silmin. Lapset ja sodan kuva vuosien 1939–1944 Suomessa* [Children and the image of war in Finland 1939–1944] (unpublished University of Tampere master's thesis, 1999).

Hietala, Marjatta, 'Adoption of Food and Meal Innovations in Finland', in: Hans Jürgen Teuteberg, Gerhard Neumann & Alois Wierlacher (eds), *Essen und kulturelle Indentität. Europäische Perspektiven* (Berlin, 1997), 322–335.

Jackson, Carlton, *Who will take our children?* (London, 1985).

Järvenpää, Taru, *Koululaisena Suursodan varjossa. Koulunkäynti, koti- ja talkootyö sekä vapaa-ajanvietto Suomessa ja Isossa-Britanniassa toisen maailmansodan aikana.* [School children in the shadow of war. Education, voluntary work and leisure time in Finland and in Great Britain during the second world war] (unpublished University of Tampere master's thesis, 1999).

Kavén Pertti, 70 000 pientä kohtaloa: Suomen sotalapset [70,000 small destinies: Finland's war children] (Keuruu, 1985).

Korppi-Tommola, Aura, (ed.), *Sotalapset* [War children] (Helsinki, 1996).

Perkins, Charles (ed.), *Childhood Memories of World War II* (New York, 1998).

Stokes, Edward, *Innocents Abroad, the Story of British Child Evacuees in Australia, 1940–1945*, (Australia, 1994).

Wicks, Ben, *No Time to Wave Goodbye* (London, 1988).

Zemon Davis, Natalie, 'Who Owns History?', in: Anne Ollila (ed.), *Historical Perspectives on Memory* (Helsinki, 1999), 19–34.

Part 3
Tradition, crisis and innovation

MARIA CONCETTA DENTONI

12 Refrigeration and the Italian meat crisis during the First World War

Conserving food, preserving it so that it does not deteriorate, is a problem which only concerns those who possess food surplus to their daily needs. Those who have little to eat on a daily basis, on the other hand, have less to conserve. This consideration alone should be enough to explain the late development and weakness of the conservation industry *par excellence*, the refrigeration industry, in Italy in the first half of the twentieth century.

Needless to say, even from the most ancient times, the beneficial effects of cooling food to preserve it were well-known in Italy. For this purpose snow and natural ice were brought from the mountains and kept until the summer in special, well-insulated pits – known as *conserve*.[1] Nor was there a lack of suitable storage in the houses of the wealthy. Otherwise, offering one's guests foods which were out of season or from distant places, was only a problem of "trusty steeds" and "well-filled purses", as the chief cook of the Gonzaga family termed it in 1655.[2] But for those who did not possess such resources, for the peasants and the greater part of the population, it had always been safer to use foods strictly linked to where they lived and to the rhythm of the seasons. This was the norm, the everyday existence of the majority, while anything else was only the 'luxury' of the few.

The industrial system in Italy was generally backward and weak (see Table 12.1) and the food industry, "closely linked to and in part submerged in the agricultural sector, which was developing at a decidedly slow rate", as Giorgio Petrocco[3] has recently noted, made little use of mechanical refrigeration or the production of artificial ice.[4] After limited, and often disastrous, attempts to import and sell frozen meat,[5] it was only when this product was adopted by the great public organizations such as the army, hospitals and charitable institutions, that imports of frozen meat began to increase. This formed the basis for the refrigeration industry which took off with the First World War.[6]

Table 12.1 *Percentage share by sectors of Italy's Gross National Product.*

	Agriculture (%)	Industry (%)
1891–1900	47.6	18.2
1901–1910	43.8	22.0

Source: ISTAT (1958), 143.

In the years preceding the war, poverty made the Italian population eat a plain diet based mainly on carbohydrates – bread, pasta or *polenta* (see Table 12.2) – or, at most, on vegetable proteins from pulses.

Table 12.2 *Food consumption in Italy per capita in kg or litres 1860–1930.*

	1861–1880	1881–1900	1901–1915	1916–1920	1921–1930
wheat	139.8	115.2	155.7	161	178
maize	46.9	33.3	33.9	26	30
paddy (rice)	12.8	10.4	13.6	19	10
legumes	13.6	11.1	15.0	13	12
potatoes	27.5	22.4	34.2	23	30
tomatoes	11.2	15.5	20.2	21	21
vegetables	36.1	39.1	51.3	61	70
fresh fruit	17.8	22.2	27.8	33	30
dried fruit	38.1	26.7	30.7	25	20
citrus fruits	8.7	9.6	12.4	16	10
beef	4.6	6.3	6.3	9	10
pork	4.8	5.6	4.9	6	5
lamb/goat's meat	1.9	2.0	1.5	1	1
other meat	3.9	3.4	3.4	4	4
fresh fish	1.8	2.5	2.7	2	3
conserved fish	1.2	1.5	1.3	1	2
milk	27.3	32.1	34.5	32	34
cheese	2.0	3.2	3.9	4	4
eggs	7.4	5.6	5.7	7	7
butter	0.5	0.7	0.9	1	1
bacon fat/lard	3.4	4.0	3.5	4	4
olive oil	8.4	5.5	5.2	5	6
sugar	2.7	2.6	4.0	5	8
coffee	0.5	0.5	0.7	1	1
wine	96.0	96.6	126.9	105	112
Average energy intakes (kilo-calories)	2,637	2,158	2,675	2,596	2,834
Food consumption as a percentage of total consumption	– –	59	58	60	54

Source: Based on V. Zamagni 1998, 176.

The consumption of meat was amongst the lowest in Europe, at around 14 or 15 kilos per year on average, which was less than half that of France and only a quarter that of Great Britain (see Table 12.3). Meat was reserved for special occasions among the poorer classes and, in any case, restricted to the parts of the animal which were not highly regarded or of low quality. In the cities, for example, where higher consumption was noticeable, meat usually meant offal, the heads or hoofs of oxen, and leftovers from the slaughterhouse. This was defined as "other meat", as opposed to beef, pork, lamb or goat's meat, or even the meat of animals slaughtered because they were no longer able to work, if not already diseased.[7] Consequently, it is not surprising that in 1914 the *Commission Scientifique Interallié du Ravitaillement* (CSIR), which was created at the outbreak of war, established in one of its preliminary studies that in Italy the percentage intake of energy derived from substances of animal origin was only 12 per cent of the total, while in France, Germany and Britain these values were more than double or almost triple this figure (27 per cent, 33 per cent and 36 per cent respectively).[8]

Table 12.3 *Average annual meat consumption per head in U.S.A, Great Britain, Germany, Switzerland and France, 1912/1913.*

Country	USA	GB	Germany	Switzerland	France
	(kg)	(kg)	(kg)	(kg)	(kg)
Meat consumption	76	57	53	42	36

Source: Albertoni & Pugliese, *Studi sull'aliamentazione* [Studies on nutrition] Napoli 1937), cit. in Zamagni (1998), 181.

On the other hand, compared with the post-Unification period (1861–1880) and, above all, the agricultural crisis of 1881–1900, the slight improvement in Italian nutrition in the early twentieth century (1900–1915) shown in Table 12.2, depended primarily on a wider use of wheat bread in place of maize in the north or minor cereals, such as the deplorable *biade*[9] in the south. The marked increase in the import of wheat can be seen in Table 12.4. Imports of wheat more than doubled in the years 1911–1913 compared with the last four years of the nineteenth century.[10]

Table 12.4 *Wheat imports to Italy, 1896–1913.*

Dates	1896	1901–1910	1911–1913
	(M quintals[11])	(M quintals)	(M quintals)
Wheat imports	6.6	10.2	13.8

Source: Porisini (1970), 512–513.

Average food consumption in Italy shown in Table 12.2 does not take into account differences between social classes. The poorer classes, especially the peasant classes, consumed less than 15 kg of meat a year. With the outbreak of war in 1915, millions of people called to arms passed from this substantially vegetarian diet[12] to one in which meat consumption increased. This change could not be sustained by the existing livestock supply.[13] In fact, the problem was that of feeding the troops employed at the front. It could certainly not be expected that the physical effort and, indeed, morale required in a wartime situation could be sustained by a diet which was poor in proteins and fats. A multitude of peasant soldiers – the backbone of the Italian army – became familiar with meat in their diet.[14] This increased consumption by about 50 per cent, compared with the 4–5 million quintals of the pre-war period.[15] This created serious problems for the future of the national livestock supply and, prospectively, for the whole agricultural economy of the country.

An attempt to reduce civilian rations did not solve the problem to any great extent, since meat was insignificant in the civilian diet, nor could a policy of reducing the quantity of meat distributed to the army be carried forward. This had been attempted briefly on the advice of two illustrious physiologists – Baglioni and Rho – who were responsible, the latter in particular, for a crusade against what was defined as "nutritional luxury" in the Italian army.[16] After the Caporetto debacle in October 1917, as can be seen in Table 12.5, the military and political hierarchy restored the old meat rations. The solution, according to the French example, would be to import frozen meat from abroad for the army and, if possible, for the civilian population. In this way the national livestock supply could be restored to normal levels of growth, without sacrificing working and milk-producing animals and bringing prices back to an acceptable level. It would also have finally been possible to bring to an end the ghastly cattle-parks, the system with which up to that time the army had been provided with meat. Besides the suffering of the cattle – left for days and days under sun or rain, with very little food, little water, often undergoing long marches to reach the soldiers' positions and subject to all manner of diseases – the controversy centred on the economic damage caused by these methods. Animals lost weight, butchering was carried out in shockingly unhygienic conditions in the open air with the loss of all those parts which could not immediately be used (hide, entrails, hooves). These were buried precariously – and were often the breeding-grounds for infection with the first hot weather. The increase in price which was caused by requisitioning of thousands and thousands of animals affected not only the livestock supply but also prices of forage, milk, and rents of grazing land.[17]

Table 12.5 *Food rations (in calories) per capita in Allied Armies, November 1917 – March 1918.*

Army	November 1917 at the front	November 1917 in base camps	March 1918 at the front	March1918 in base camps
Italian	2,917[1)]	2,564	3,303.0[1)]	2,802.0
French	3,424	3,156	2,926.5[2)]	2,673.0[2)]
British	4,400	3,600	3,829.5[1)]	3,326.0[1)]
American	3,800	3,250	3,781.9	3,213.0
Belgian	n.a.	n.a.	3,658.9[1)]	3,658.9

Source: Giuffrida & Pietra (1936), Appendix, 330.

Note: [1)] without supplement
[2)] without money

In the years preceding the conflict, the War Ministry had attempted to introduce frozen meat among the troops. Indeed, in 1911 a five-year contract had been signed with the Union of Cattle Importers, the company which also supplied the army with meat on the hoof, that is live cattle, that came generally from the Balkans. Upon the outbreak of the European War in August 1914, the army was placed on a state of alert – an obligatory measure, in spite of Italy's declaration of neutrality. Its size was increased by about two-thirds. It became clear that it was not possible to provide food for the army in the usual way. This was not only because the quantity of meat available would not have been enough, but because the state of other nations at war and the requisition of British shipping in particular,[18] prevented the arrival of frozen meat and imported animals. It was thus necessary to requisition from the countryside. In this way, during Italy's ten months of neutrality, enormous damage was done to the livestock supply, and particularly cattle. An imperfect knowledge of national statistics, which were still based on the 1908 census, allowed harmful optimism to persist. This postponed the radical decisions necessary to provide a solution to the problem.

Even after Italy's entry into the war alongside France, Russia and Britain, requests made by Italy for imports of wheat and frozen meat through the United Inter-Allied Front – were below real needs. This created an extremely embarrassing situation which, furthermore, was also economically ruinous. In 1917, for example, in order to hide the reality of a disastrous wheat harvest, some 10 million quintals below the less than satisfactory previous year, the Agriculture Minister, Raineri, released triumphant declarations like "Italy has no need of a bread ration card, our grain situation is superior to any other state, be it neutral or allied."[19] Such claims, as he later tried to justify himself, above all with the British, were made in order to subdue internal worries.

With regard to meat, conditions were perhaps even worse. Experts in the sector were accused of scaremongering[20] and the figures put forward by the director of logistical services of the War Ministry, General Tettoni, were not accepted. He had estimated a need to import 20,000 tons of meat a month –

the same as France, which had a greater supply of livestock in the country and in the colonies. These imports, as the General explained, would be sufficient for the army, for the army slaughterhouses, where reserve rations (tins of preserved meat) were prepared and for the civilian population. Thus, this represented the total quantity of Italy's requirements. However, it was agreed to seek a maximum quota of 7,000 tons a month, subject to review every six months. For the first year, imports averaged 5,200 tons a month. This quantity was not only less than Italy needed but, even more absurdly, was less than what Britain had declared it was willing to provide.[21]

At the root of this self-limitation of meat requirements, was the economic concern that frozen meat prices had increased by 100 per cent since 1911.[22] In addition, there was a whole series of other difficulties, in part a consequence of the war, in part the result of governmental uncertainty and incapacity. First of all, at the root of all considerations, was the conviction that the war would last only a short time – three or four months at most. There was therefore no forward planning for food supplies which were left to improvisation.[23] As far as the consumption of frozen meat is concerned, once the impossibility of repeating requisitions *ad infinitum* was realized, doubt persisted as to whether imports should be solely for the army or also for the civilian population. This left unsettled the quantities required, or which Ministry was responsible. Once the supply of imported meat to Italy had been agreed, for which Britain bore the total costs, it was expected that the pooling of all allied refrigerator ships would occur (see Table 12.6).[24] In Italy it was therefore necessary to provide 'cold' spaces in which to store shipments in the ports of Naples and Genoa. What was even more of a problem, was the necessity to distribute consignments of frozen meat to the front and behind the lines by thermally insulated trains and lorries. At the beginning of the war this type of facility was inadequate. There were only 250 refrigeration systems, with 18,100 tons capacity, 12,700 of which were available to the military administration. Only 60 highly insulated wagons, suitable for meat transportation by rail were available and absolutely nothing like refrigerated lorries. The War Ministry set to work immediately, to solve the problem and to meet its own requirements. On the other hand, the situation was completely different with regard to the storage and distribution of meat for the civilian population.[25] Without having made any previous provisions in this respect, responsibilities were divided up between the various ministries: War, Agriculture, Treasury and the Interior. The latter, for example, in its Health department, was responsible for the management of veterinary problems. This was seen only as a question of preventing the transmission of diseases from animals to humans. It was not seen as an economic problem in relation to animal breeding or of animals useful to agriculture.[26] It was not therefore the responsibility of the Agriculture Ministry, which had not existed autonomously prior to June 1916.

Table 12.6 *Allied refrigerator ships during the war.*

	Britain		Italy		France	
At the beginning of the year	No. of steamships	Cargo capacity per journey (tons)	No. of steamships	Cargo capacity per journey (tons)	No.of steamships	Cargo capacity per journey (tons)
1914	209	400,000	--	--	--	--
1915	211	405,000	--	--	--	--
1916	291	500,000	8	18,200	--	--
1917	323	630,000	9	19,700	11	21,500
1918	283	507,000	9	19,700	11	21,500
1919	229	450,000	10	20,350	11	21,500
1920	229	450,000	10	20,350	16	31,000

Source: Zingali (1920), 270.

If we add to this situation the continuous interference and vetoes reflecting Treasury concerns about the increasing costs of the war, it can be understood what effect this had on the nutritional situation. The full weight of the restrictions fell upon the weakest sector of the country, that is to say the civilian population which was not immediately employed at the front. For example, there was the attempt to reduce the amount of grain imported, but avoid the introduction of ration cards – which it was feared would have dire consequences on the morale of the population and consequently also influence the troops at the front. However, bread regulations were introduced. By decree normal standard bread was produced from sifted flour of an extraction rate of 80 per cent instead of 70 per cent, as was usual. This regulation, imposed in March 1915 two months before Italy's entry into the war, was the only preventive measure taken. A whole series of other decrees between March 1916 and February 1917 modified the loaf even further: loaves were to be 700 grams, the extraction rate was raised to 90 per cent, and bread was to be sold stale. Flour substitutes of every type were introduced: as well as bran, there was also flour made from rice, maize, chestnuts, and lupin seeds in order to make a saving of 10 to 15 kg of flour a quintal. These measures were to the detriment of the consumer, who paid for 10 to 15 kg of water and ate bad bread, made from inferior grain.[27]

Once consumption of fresh meat had been reduced by 50 per cent, it was intended to distribute only frozen meat among the civilian population. With this aim[28] (see Table 12.7) the War Ministry offered municipal administrations the opportunity to buy some consignments of frozen meat in January 1916. They received, however, very little interest. What factors influenced this refusal? First and foremost, the purchases were irregular, as the War Ministry could not guarantee a continuous supply. At the port of Naples, for example, some months after this offer, there would have been two full months from 7 April 1916 to the first half of June, without the arrival of any meat. In Genoa, from 7 May to the first part of June in the same year the same situation would have occurred.[29] In these circumstances, how could local administrations be expected to invest in the extremely expensive equipment necessary for the conservation and sale of frozen meat? On 30 November 1915, according to statistics of the Health Ministry, only 46 of the 69 provinces of the Kingdom were provided with refrigeration systems.[30] These were distributed among 150 municipalities – two-thirds of which were in Piedmont and Lombardy. Of the total number of 250 systems only 40 were installed in municipal slaughterhouses. For an immediate comparison, we only need to consider that in Germany in the same period there were 800 municipalities with slaughterhouses equipped with refrigerators.[31] The data in Table 12.8 clearly show that Italy's refrigeration potential during the war was inferior both to that of Great Britain and France. Other obstacles stemmed from the high cost of frozen meat at two lire a kilogram in 1915 against 2.12 lire for fresh meat, or 2.10 lire in 1916 against three lire for fresh meat. By comparison, in Britain frozen meat cost the equivalent of 1.25 to 1.60 lire per kilo, according to the cut and the quality.[32] Above all, even the poorest classes regarded frozen meat with absolute perplexity, not to mention disgust.

Table 12.7 *Imports of frozen meat by the allies.*

Years	Britain (tons)	Italy (tons)	Other Allies (tons)	Total (tons)
1913	720,000	9,117	36,883	766,000
1914	695,000	3,312	102,688	801,000
1915	662,000	29,171	190,829	882,000
1916	534,000	90,989	291,011	916,000
1917	433,000	100,695	431,305	965,000
1918	489,000	153,459	487,541	1,130,000
Percentage of total imported				
	(%)	(%)	(%)	(%)
1913	94.0	1.2	4.8	100
1914	86.8	0.4	12.8	100
1915	75.1	3.3	21.6	100
1916	58.3	9.9	31.8	100
1917	44.9	10.4	44.7	100
1918	43.3	13.6	43.1	100

Source: Zingali, cit. in Bachi (1926), 605.

Table 12.8 *Allied frozen storage capacity during the war.*

	Britain		Italy		France	
Year	Cubic capacity (1000m^3)	Capacity (1000 tons)	Cubic capacity (1000m^3)	Capacity (1000 tons)	Cubic capacity (1000m^3)	Quantity (1000 tons)
1914	850	2,200	n.a.	n.a.	n.a.	n.a.
1915	1,000	2,600	157	18.1	15	17
1916	1,000	2,600	170	19.0	323	36
1917	1,000	2,600	182	21.0	390	43
1918	1,220	3,150	248	27.7	448	50
1919	1,500	3,700	270	30.0	540	60

Source: Zingali (1920), 278.

'Refrigeration theorists', more or less in good faith, put this down to the ignorance of the masses. They also bitterly criticized the behaviour of the government which, while protesting the need not to ruin the livestock supply, continued during the war to levy import duty of 120 lire a ton on frozen meat. Subsequently the duty was raised in proportion to the increase in the price of the meat; by 1916, it amounted to 160 lire a ton. In addition, the duty on the jute and the cotton wrappings of frozen beef quarters totalled 4.80 lire for every ton of meat. Finally, the health tax, still a government right, came to 145.80 lire a ton, which meant 15 cents more on the sale of a kilogram of meat.[33] Why should the government impose a duty to defend Italy's livestock industry during the war years, when it was essential to increase to the maximum the import of frozen meat in order not to destroy the livestock supply? Uberto Ferretti, one of the pioneers of the cold storage industry in Italy and also the founder of the *Refrigeration Review*, described the duty on frozen meat as "antidemocratic".[34] He called for the duty to be abolished at all costs, as had happened in France, by the decree of 2 August 1914. If the price had been substantially lower, argued Ferretti and the other refrigeration experts, not simply a few cents – which among other things was immediately lost by the decrease in weight due to the defrosting process – more faith would have been placed in frozen meat. It would then have been appreciated as much, if not more than fresh meat, because it was better hung.

In reality, and even Ferretti recognised this, frozen meat in Italy had always been sold badly, without the correct defrosting process, for which there was neither sufficient equipment nor technical expertise. Worse still, on the part of the tradesman, there was always the dishonest aim of burdening the *sfriso*, that is the loss of weight that the meat underwent, on the consumer. Even at the moment of buying meat, in the countries of production, wrong – if not dishonest – choices were frequently taken. With the pretext of taking into consideration the tastes of Italians – no great lovers of fatty meat – batches of emaciated animals were bought, of significantly inferior quality but, obviously at heavily reduced prices. The result was high profits for the importer but bad returns for the buyer.

The lack of knowledge with regard to defrosting had also caused problems in the use of frozen meat by the army, in the years preceding the war. A nauseating broth was distributed to the soldiers, with globules of fat on the surface and boiled meat which was not only tough and with a very unappetising appearance, but above all, according to Ferretti "insipid and of no nutritional value".[35] It was not uncommon in purchases for the civilian population to find batches of meat which were too old, if not in part already rotten. Commonly, this resulted from transporting it by ships which were not suitable to carry frozen meat. To all these problems a remedy could have been found, however, according to the experts in refrigeration science, by importing meat in large quantities for immediate sale, instead of leaving it for long periods in refrigerators. The provision of suitable instructions for defrosting and cooking was also necessary for its acceptance.

By winning this battle for progress and improving the country's nutrition, Italy's livestock supply would have been saved. At the same time a new industry would have been created which would have been wholly Italian, including the machinery and patents, without having to rely on Germany any longer, even though, it should be emphazised, that by the end of the nineteenth century some Italian engineering companies had already begun to construct refrigeration systems – the Fonderia Pignone of Florence, which by 1909 published its own catalogue *Le macchine frigorifiche* (Refrigeration Plants)[36], the Fonderia Gaetano Barbieri and Co of Castel Maggiore in the province of Bologna and the Workshop of Otto Sigg and Co of Turin (later to become the Fabbrica Italiana di Torino).[37]

It was the economic implications of developing new industries more than humanitarian questions – or improving nutrition – that aroused the interest of refrigeration manufacturers. Meeting in Rome in December 1916, for the first time during the war, the refrigeration manufacturers pressed the Government strongly to introduce "an honest and proper protection".[38] In addition to the protests against government and municipal duties for imported frozen meat, the refrigeration industry requested concessions to buy sodium chloride for refrigeration use at the same price as the more favoured industries. Soap factories, for example, obtained salt at 7 lire a quintal not at 25 lire, as the Finance Ministry required on the basis of old tariffs. These tariffs grouped the refrigeration industry's freezing capacity along with the production of ice cream and sorbet, which were considered luxury items.

Pressed by the needs of the war, the increase in the price of fresh meat and the need to produce working animals, the Government demonstrated increased attention towards the problem of refrigeration, granting both the request for sodium chloride, at 10 lire a quintal, and also other concessions asked for at the manufacturers' conference. In 1918, after obtaining increased supplies of meat from America, it was the new director for Supplies and Consumption, the industrialist Silvio Crespi, who issued a series of decrees for the creation of a true national refrigeration system. Thus firms were provided with tax relief, private property expropriated by the state was sold at low cost for the construction of refrigeration plants and the authorities gave precedence to businesses presenting requests or projects regarding this area.[39] The usefulness of these provisions, issued on 21 November 1918, even if too late for the war which was already concluded, signalled a positive development for the refrigeration industry and, in general, better nutrition for the people. In 1915 there were only 250 refrigeration plants in Italy. By 1918 the refrigeration industry had expanded to 1,400 and by 1922 had more than doubled to 3,000 plants. There is no doubt that the growth of the food preservation industry in Italy was strongly influenced by the First World War.[40]

NOTES

1. Petrocco (1998), 390–391, 417–419.
2. Montanari (1993), 203, 204; see also Aymard (1975), 431–444.
3. Petrocco (1998), 419.
4. Brignone (1993), 88, 91–95.
5. For the attempts of Fausto Scerno, a Genoese importer, see Ferretti (1913), 17.
6. Petrocco (1998), 424.
7. Dentoni (1995), 71, 75; see also Table 12.2. For diseased animals the popular proverb quoted in Sorcinelli (1983), 94, is significant in this regard: "Se un povero mangia una gallina o è malato il povero o è malata la gallina" [If a poor man eats a chicken, either the poor man or the chicken is ill].
8. Sorcinelli (1983), 94–95.
9. *Biade*, a pejorative term for the poorest quality cereals only suitable for animals.
10. Zamagni (1998), 177; see also Tables 12.2 and 12.4.
11. For the purposes of this chapter reference is made to the metric quintal of 100 kg (or 220 lbs).
12. Bachi (1926), 423–424 refers to the peasant population as "abitualmente vegetariani" [habitually vegetarian].
13. Italian livestock supply according to the Census of 1908:

Horses	955,878
Donkeys	849,723
Mules and hinnies	388,337
Cattle	6,198,861
Swine	2,507,798
Ovines	11,162,926
Goats	2,714,878

From Fotticchia, N., 'La produzione zootecnica italiana' [Italian livestock production], *Nuovi Annali del Ministero per l'Agricoltura* (1922), quoted in Bachi (1926), 423. For imports and exports see Bachi (1926), 46.

14. According to the 1911 census 55 per cent of the active Italian population was employed in agriculture. Factory workers were in reserved occupations during the war period.
15. The data on meat consumption in the army are not consistent. See, however, Zingali in Bachi (1926), 590 and following pages; according to ISTAT (1958), an average availability of 5,128,000 quintals is reported for the period 1901–1910, against an average of 7,017,000 quintals for the four years of the war.
16. Zingali in Bachi (1926), 530 and following pages.
17. Ferretti, 'L'industria del freddo e la difesa nazionale' [The refrigeration industry and national defence]. *Rivista del freddo* 8 (1916), 230–232; Zingali (1920), 267–268.

18. Ferretti, 'La guerra europea e l'industria del freddo' [The European war and the refrigeration industry]. *Rivista del freddo* 1 (1915), 6.
19. ACS, *Presidenza Consiglio dei Ministri* (PCM) 17.1.8, dossier 2, Mayor's reports to Sonnino, 29 and 30 June 1917, n. 1073/640 and n. 1075/641; ibid., *Carte Orlando*, b. 65, dossier Affari finanziari, Corrispondenza tra Ranieri e Orlando [Financial affairs, correspondence between Ranieri and Orlando], 7 May 1917.
20. Among the most critical Ranieri, not yet Minister for Agriculture, who in *La Venezia agricola* defined the meat problem a "fearful point" starting, however, from the assumption that the war would only have lasted six months, with the mobilization of two million men, cf. *Rivista del freddo* 7 (1917), 274–275. Mobilization would later have involved more than three million men, for a period which was decidedly longer than that hypothesized by Ranieri.
21. Ferretti, 'L'industria del freddo'. *Rivista del Freddo* 8 (1916), 239, 244, 246.
22. Zingali (1920), 282–288.
23. Dentoni (1995), 15–33.
24. Ferretti, 'L'impianto del piroscafo frigorifero italiano *Resurrezione*' [The refrigeration system of the Italian refrigerator steamship *Resurrezione*]. *Rivista del freddo* 12 (1916), 354–360. It was expected by the British that pooling would include the steamship *Resurrezione*, bought and fitted out by the local administration of Milan to provide supplies for the population.
25. Cattaneo, 'I mezzi per i trasporti frigoriferi in Italia' [Refrigerated transportation in Italy]. *Rivista del freddo* 1 (1920), 22–24; Ferretti, 'Il trasporto delle carni congelate' [The transportation of frozen meat]. *Rivista del freddo* 11 (1915), 416–427.
26. Ferretti, 'Lettera aperta a S.E. il Ministro di Agricoltura' [Open letter to His Excellency the Minister for Agriculture]. *Rivista del freddo* 12 (1916), 161–162.
27. Dentoni (1995), 37–41 and more widely, Dentoni (1991), 133–147.
28. Zingali (1920), 265, Note 1; Puglisi, A., 'Il problema zootecnico nel momento attuale e nel dopoguerra' [The livestock problem at present and in the post-war period]. *Rivista del freddo* 10 (1917), 236. The government succeeded in obtaining from Britain a promise for an increase in imports which should have become 10,000 tons monthly, after September 1916, but in reality never exceeded 6,000 tons a month.
29. Ferretti, 'L'industria del freddo'. *Rivista del Freddo* 8 (1916), 246.
30. Ministero dell'Interno, Direzione Generale della Sanità pubblica (1916), 279. In reality the refrigeration plants which were taken into consideration were 209, neglecting a further 41 plants because they were "piccolissimi" [very small] according to the above mentioned administrative department for Public Health.
31. Ferretti (1913), 133.
32. Zingali (1920), 282–296.
33. See also the observations made by G. Griglio, Director of the Public Slaughterhouse in Palermo, in: 'Per l'abolizione del dazio doganale' [For the abolition of import duty]. *Rivista del freddo* 2 (1916), 19–20.
34. On import duty as "antidemocratic", see the observations made by Ferretti in reply to Griglio, (Note 35 below), and also Ferretti, 'La seconda requisizione dei bovini' [The second cattle requisition]. *Rivista del freddo* 5 (1916), 121.
35. "Insipida e esaurita". Ferretti, 'Il commercio e il consumo delle carni congelati in Italia' [The commerce and consumption of frozen meats in Italy]. *Rivista del freddo* 1 (1915), 34; id., ibid. 2 (1915), 55–57.
36. Ferretti, 'L'industria del freddo e le sue applicazioni all'igiene agricoli e commerci' [The refrigeration industry and its applications in agricultural and commercial hygiene]. *Rocca San Casciano* (1909), 41–43.
37. With the war, refrigeration businesses, even the most simple such as ice factories, underwent a significant growth, above all if declared "auxiliary establishments" (as *Gondrand Mangili* or the *Ghiacciaie e Neviere Naploitane*); in this regard the documentation contained in ACS, *Ministero delle Armi e Munizioni* [the Ministry for Arms and Ammunition], folders n. 6, 13, 14, 20. As regards the manufacturers of refrigerators, one of their requests was, in particular, protection against foreign

industries; against the total abolition of duties on foreign refrigeration systems, as had been proposed by A. Giraldoni, cf. Giraldoni, 'L'alimentazione carnea e l'intervento dello stato' [Meat nutrition and state intervention]. *Rivista del freddo* 2 (1916), 38) in June 1916. Ferretti considered it not yet possible in Italy to provide everything for the refrigeration industry; only the *Fonderia del Pignone* and the *Ditta Barbieri* would have been able, according to his claim, to construct all the refrigeration systems with their own means, cf. Ferretti, 'Per la tutela dell'industria nazionale' [For the protection of the national industry]. *Rivista del freddo* 6 (1916), 165.

38. Cf. 'Atti del 1° Congresso nazionale degli Industriali del Freddo, Roma 15/16 dicembre 1916' [Acts of the First National Congress of Refrigeration Industrialists, Rome 15/16 December 1916]. *Rivista del freddo* 12 (1917); see also ACS, *Ministero dell'Interno, Direzione Generale della Sanità pubblica*, Atti amministrativi 1910–1920, b. 91, 17 October 1916, Ferretti's request to Lutrario, the general director of the public health service, for a financial contribution which was later granted at £200, ibid., Health Administration to the Under-secretary for Internal Affairs, 28 October 1916, n. 20173.C/40535.
39. Ferretti, 'I provvedimenti del Governo' [Government provisions]. *Rivista del freddo* 11 (1918), 308–310 and id. 'L'azione del governo' [Government action]. *Rivista del freddo* 12 (1918), 313–318, where the text of the Decree of the 21 November 1918, n. 1747. On Crespi's action see also Dentoni (1995), 125.
40. Cf. C. Corbetta, 'Lo sviluppo delle industrie frigorifere in Italia' [The development of the refrigeration industry in Italy]. *Rivista del Freddo* 12 (1919), 401 and Petrocco (1998), 426.

BIBLIOGRAPHY

Archivio Centrale di Stato [Central State Archives], Rome (ACS):
Carte Orlando [Orlando Papers].
Ministero Armi e Munizioni [Ministry for Arms and Ammunition].
Ministero dell'Interno, Direzione Generale Sanità Pubblica [Ministry for Home Affairs, Administrative Department for Public Health].
Presidenza Consiglio dei Ministri [Presidency of the Cabinet of Ministers].

Aymard, M., 'Pour l'histoire de l'alimentation: quelques remarques de méthode'. *Annales ESC* 2–3 (1975), 431–444.

Bachi, R., *L'alimentazione e la politica annonaria in Italia* [Nutrition and food policies in Italy] (Bari, 1926).

Brignone, D., 'Le società riunite Fabbrica di giaccio e Ditta Francesco Peroni: nascita e sviluppo dell'Industria del freddo a Roma a cavallo del secolo' [The united companies *Fabbrica di giaccio* and the *Ditta Francesco Peroni*: birth of the refrigeration industry in Rome in the early century], in: Brignone, D. (ed.), *Innovazione tecnologica ed Industria in Italia* (Rome, 1993), 87–126.

Dentoni, M. C., *Annona e Consenso in Italia, 1914/1919* [Food and agreement in Italy, 1914/1919] (Milan, 1995).

Dentoni, M. C. '"L'arte di vivere bene mangiando poco". Signore e contadine di fronte ai problemi alimentari' ["The art of living well while eating little". Women and female peasants dealing with problems of nutrition]. *Annali Cervi* XIII (Bologna, 1991), 133–147.

Ferretti, U., *L'industria del freddo e le sue applicazioni* [The refrigeration industry and its applications], 2nd edn with Preface by Ch. Tellier (Arpino, 1913).

Giuffrida, V. & Pietra, G., *Provital. Gli approvvigionamenti alimentari durante la guerra* [Food supplies during the war] (Padua, 1936).

ISTAT [Italian Central Statistics Institute], *Sommario di statistiche italiane 1861–1955* [Summary of Italian statistics 1861–1955] (Roma, 1958).

Ministero dell'Interno, Direzione Generale della Sanità pubblica [Ministry for Home Affairs, Administrative Department for Public Health], *Notizie statistiche sui frigoriferi esistenti in Italia al 30 novembre 1915 per la conservazione delle carni fresche o congelate e degli altri prodotti alimentari di origine animale* [Statistics on existing refrigerators in Italy on 30 November 1915 for the conservation of fresh and frozen meats and other animal based food products], 1916.

Montanari, M., *La Fame e l'abbondanza* [Hunger and abundance] (Roma / Bari, 1993).

Petrocco, G., 'La conservazione del cibo: dal sale all'industria agro-alimentare' [The conservation of food: from salt to the agro-alimetary industry], in: *Storia d'Italia, Annali XIII, L'alimentazione* (Turin, 1998), 379–447.

Porisini, G., 'Produzione e produttività del frumento in Italia durante l'età giolittiana' [Production and productivity of wheat in Italy during the Giolittian period]. *Quaderni storici* 14 (1970), 507–540.

Rivista del Freddo [Refrigeration Review], 1915–1919.

Sorcinelli, P., 'Note sull'alimentazione nell'età giolittiana' [Notes on nutrition in the Giolittian period]. *Italia Contemporanea* 150 (March 1983), 89–94.

Zamagni, V., 'L'evoluzione dei consumi fra tradizione e innovazione' [The evolution of consumption between tradition and innovation], in: *Storia d'Italia, Annali XIII, L'alimentazione* (Turin, 1998), 171–204.

Zingali, G., 'Gli aspetti economico-statistici del problema della carne congelata' [Economico-statistical aspects of the problem of frozen meat]. *La Riforma Sociale* XXXI (7/8) (1920), 261–296.

Zingali, G., 'Il rifornimento dei viveri dell'esercito italiano durante la guerra' [The supply of provisions to the Italian army during the war], cit. in Appendix to Bachi, R., *L'alimentazione e la politica annonaria in Italia* [Nutrition and food policies in Italy] (Bari, 1926), 513–647.

HEATHER HOLMES

13 Food from the hedgerows

Collecting wild fruits and plants in Scotland during the Second World War and its aftermath

During the Second World War Scotland experienced great changes in food production. The area of agricultural land under cultivation increased, not only by the ploughing up of grassland but also by the supplementary use of land in towns. In urban centres such as Edinburgh, Glasgow and Dundee, as well as in smaller towns, areas of grass in school grounds and parks were brought into crop production. The cultivation of allotments was expanded and householders were encouraged to grow vegetables and fruit in their gardens. Stress was also placed on improvements in husbandry. On land under crops, emphasis was placed on increasing yields. Greater use was made of wild and uncultivated plants and fruits from wastelands. These changes to food production lasted for some years after the war and altered patterns of food consumption in Scotland.

This chapter looks at aspects of the relationship between rural and urban areas as production, processing and consumption centres in Scotland during the Second World War and the years immediately following it. This is examined through a case study of the collecting and utilization of uncultivated fruits and plants which could be secured from a range of habitats, the most important being hedgerows. The following account will discuss the relations that existed between these areas in the work of collecting, processing and utilizing these fruits and plants.

Role of wild fruits and plants before the war

Before the outbreak of the Second World War a wide range of wild fruits and plants were gathered. These included brambles (or blackberries), crab apples, haws (hawthorn berries), rosehips (fruit of the wild rose), elderberries, rowans, sloes, and cranberries. Some of them, such as brambles and rosehips, which grew in profusion over a wide area of Scotland, were gathered on an extensive scale. Their fruit was widely utilized and was "fairly fully harvested in any normal year".[1] Locally, and nationally, its use depended on availability, yields, and the resourcefulness of gatherers to secure it.

Fruit was primarily used for household purposes. It could be eaten raw, or used as an ingredient in puddings and other dishes. More frequently, it was processed into preserves (jams and jellies), bottled or made into chutneys and ketchups; some was made into wines and spirits.[2] The fruit and berries and their processed products had a number of roles. Both acted as a major source of fruit for the household; where cultivated fruit was also available, it could augment its supply. Plant matter from stems, leaves, fruits and roots, was utilized for a range of medicinal purposes such as herbal remedies, drugs, preparations and welfare foods.[3]

Schemes for wild fruits and plants

Shortly after the outbreak of the Second World War, government departments in Scotland, as in England and Wales, recognized that rural areas formed an important resource which could assist in the national war effort. Rural areas already had a well-established tradition of collecting and utilizing wild fruits and plants which could be more fully exploited through systematic organization. Produce from the wild could be utilized for a number of uses: fruits and plants could counteract shortfalls in production caused by poor yields, and breach the gap between supply and dearth in cultivated crops. It could also provide a range of substitute products to counteract war-time shortages. To utilize the wild fruits and plants growing in rural areas, government departments in Scotland instituted a number of schemes to organize their collection, processing and distribution for consumption. Three such schemes were introduced, all of which formed part of government plans to increase food production.

The first of the schemes, which dealt with wild fruit and plants, aimed to collect brambles, "and other wild fruits" specifically "bilberries (blaeberries), cranberries, elderberries, rowan berries, crab apples".[4] It had developed from an earlier scheme implemented in 1918, at the end of the First World War, when "the comparative failure of the country's fruit crop" required that fruit was obtained from all available sources, so that a shortfall could be averted in the preserves' ration for the forces and the civilian population.[5] It was not until 1941, however, that a similar situation arose during the Second World War: cultivated and stone-fruit harvests were poor, so that it became essential for collection to be instituted.[6] From that year until the 1950 harvest,[7] fruit from wild plants was gathered on a systematic basis throughout Scotland. Its collection allowed for the maintenance of the preserves' ration.

In the second scheme, the Scottish Medicinal Plant Scheme, plants were gathered for medical purposes such as drugs, preparations and welfare foods. Like the bramble scheme, it also had been instituted during the First World War.[8] It, too, was not immediately introduced at the start of the Second World War and did not commence on a national basis until 1941, the year following the introduction of the Welfare Foods Service.[9] Under Medicinal Plant Scheme, a wide range of plant materials was collected. At the start of the scheme these were described as "leaves and roots of common plants". They included sphagnum moss, stinging nettle, common foxglove, dandelion, raspberry, elder, male fern, valerian, colts foot, burdock, and rosehips.

However, their extent became more limited: from 1943, and until the end of the war, only rosehips and foxglove leaves were gathered.[10] Thereafter until 1952, only rosehips, were collected; nationally, rosehips continued to be gathered until at least 1958.[11] Of all plants and fruits, rosehips, a rich source of the protective vitamin C, made the single most important contribution to the scheme.

In the third scheme, surplus fruit was processed into preserves in organized fruit-preserving centres located throughout Scotland. Unlike the first two schemes that focused on the collection and utilization of wild fruits and plants, this one was primarily developed to deal with surplusses of cultivated fruit grown in gardens and allotments in rural and urban areas alike; wild plants formed one aspect of that wider work. Its role was not primarily to increase the total output of jam, but rather act "as a means of preserving the wastage of fruit which cannot otherwise be consumed".[12] It also had one further difference. Like other schemes that dealt with the cultivation of crops on agricultural land or in urban areas, it came into operation before those which were concerned with resources from nature.[13] Preserving centres, such as found in schools and hospitals, operated from the 1940 season onwards until the end of the war,[14] though schools continued that work until 1950.[15] Such centres acted as focal points where surplus fruit could be brought to be processed and disposed of.

Rural and urban partnerships

The operation of the three schemes involved a number of partnerships between people in rural and urban areas, gatherers and preservers, and suppliers and consumers. All three schemes were centrally organized by Scottish government departments located in Edinburgh. As these departments were in an urban area away from where wild fruits and plants grew in abundance, they instituted and developed an extensive network system to link their central organization with collecting areas throughout Scotland. That network was facilitated through a number of partnerships, which were shaped by the work that was to be undertaken and the plants and fruits to be dealt with. To arrange the collection of all wild fruits and plants, the most important partnership was between the Scottish Education Department, the department responsible for the administration of education in Scotland, and Local Education Authorities (some 38 in number throughout the country) and their schools, which supplied the local organization for collection and some of the processing work.[16]

Collection work embraced a range of geographical areas across Scotland, the bulk of which was undertaken in rural counties and districts. The Scottish Education Department noted in 1941 that it had been asked to write to its Local Education Authorities "to bring this need to the notice of Head Teachers of country schools, so that their pupils may help with this part of the war effort".[17] But the Department also issued collection instructions to urban authorities, such as Edinburgh and Glasgow, as well as to heavily industrialized areas. Glasgow was the most important of these, and its schools participated in the work between 1943 and 1945.[18] Urban areas also assisted in the work of gathering rosehips. When that work was restricted in extent during 1950, it nevertheless still included Glasgow,[19] and in the following two years when it was again extended, it covered all the major urban areas: Edinburgh, Dundee and Glasgow.[20] Gatherers came from a range of backgrounds, including farms; while children from villages and towns went out to their neighbouring countryside.[21]

This educational partnership was essential in enabling collection activities to be undertaken. It allowed a large labour force to be mobilized to undertake the most labour intensive part of two of the three schemes. Although no statistics exist of the numbers involved, evidence suggests that school children and young persons, including members of youth organizations, made important contributions to the work, as they also did for other wartime schemes such as the collection of waste paper, metal and metal foils, bones for recycling, and harvesting activities. So important was that assistance that the Scottish Education Department reported that the bulk of the work of collecting rosehips during the war years was "largely dependent on the assistance" given by children and young people.[22]

Other partnerships were established to undertake different aspects of the activities in the schemes. These were important for dealing with the collected matter and preparing it for processing. Again, they joined urban and rural areas, as well as the institutions and community structures within them. As with the arrangements for collection work, these arrangements involved a partnership between Scottish government departments, notably the Scottish Education Department, and local organizers through two voluntary organizations, the Scottish Women's Rural Institute (SWRI) and the Scottish Women's Voluntary Service (SWVS), both of which had their headquarters in Edinburgh.[23] Both had networks throughout Scotland that linked their central organization with their local branches. At a local level, they worked in partnership with schools and other collecting centres: schools acted as centres where gathered material was held and made ready for dispatch to the voluntary organizations or to other processors. Although some schools and collecting centres had facilities to dry the plant matter, not all did, and a close partnership developed between the schools and the local representatives of the women's voluntary organizations so that such work could be undertaken.

These local and regional organizations collaborated with manufacturers located in urban areas. All rosehips, for example, were conveyed to a central location in the industrial town of Airdrie, in Lanarkshire, from which the dried plant matter was sent to be utilized for medicinal use. Processing enabled the rosehips, which had hairs that had to be removed using specialist equipment, to be manufactured into syrup. During the war years, processing was undertaken by one central manufacturer in Edinburgh, Messrs Duncan, Flockhart and Company of Holyrood Road.[24] When arrangements altered after that period,[25] fruit secured in southern counties was processed by a firm based in Newcastle-on-Tyne, Messrs Scott and Turner Ltd.[26]

Not all of the schemes, however, required the collected matter to be processed by a central manufacturer or manufacturers. Where fruit was to be preserved, that work was undertaken on a local basis, which involved collaboration between the Local Education Authorities and other institutions. However, national directives were required from the Scottish Education Department to set them up and ensure that they had a strict code of practice and operation. A further partnership was also established between these centres and the Ministry of Food which allowed them to obtain a supply of sugar for preserving purposes. As the school log book for Spott, a small rural school in East Lothian, notes, "permission has been obtained from the Ministry of Food to obtain 28 pounds of sugar to be used in making jam for use in the school".[27]

The task of preserving fruit thus linked rural and urban areas, and resource areas with processing centres. As these centres primarily dealt with surplus produce from gardens and allotments, they were located in both rural areas and towns. During 1942, in East Lothian, a predominantly rural county, centres were established in all the major villages and towns: Haddington (the county town), Longniddry, Cockenzie, Aberlady, Gullane, Dunbar, Saltoun and Gifford, and East Linton.[28] Others, such as hospitals, orphanages and communal feeding centres, were primarily located in urban areas. The task of preserving fruit in these centres established further partnerships, and linked gatherers who brought their fruit to be preserved and processors. Local members of the SWRI assisted in the activity of fruit-preserving, as did teachers, who also acted as demonstrators for the work.

Directing the schemes

These partnerships were central to the work and success of the three schemes. However, they could not satisfactorily operate without publicity to promote them. Promotion material linked all the levels of work and the rural and urban areas and resource-collecting and fruit-processing centres. Publicity material was provided at a number of levels. The most important was issued by government departments in their official correspondence which linked central and local organizations and rural and urban areas. The Scottish Education Department issued circulars and memoranda to all Education Authorities which outlined the schemes and how they would work; in turn, those documents were sent to school managers. In their press releases, other government departments, such as the Ministry of Food, linked urban and rural aspects of the schemes. In one press notice, *Fruit Preservation Plans for 1943*, which dealt with the preservation of surplus fruits, the Ministry of Food asserted how it "is once again appealing to the public in both town and country to support the scheme for co-operative fruit preservation".[29]

The schemes were also reported in a range of media that brought aspects of the work to the attention of the general public. Most important was the national newspaper, the *Scotsman,* and the largest provincial newspaper, the *Glasgow Herald*; both were based in urban centres. In rural areas, that function was served by the national agricultural and rural press titles, such as the *Scottish Farmer* and the *North British Agriculturist*. Both urban and rural presses utilized their 'household columns' as an important vehicle for transmitting information: they informed women (who looked after the household and the food resources within it) of the schemes, the need to undertake the collection work, and the methods to be used for handling and preserving material; they also reported the success of the work.

The urban directives, especially the circulars and memorandum issued by government departments, played a central role in the operation of the schemes. Those issued by the Scottish Education Department to its Education Authorities performed a number of roles. They ensured that all activities were systematically undertaken: they aimed to maximize fruit and plant collection, and to make the best possible use of the fruits and plant materials and labour resources. Furthermore, they ensured that the gathered produce was of a high quality and that the collected plant material was dealt with quickly and efficiently.

The Scottish Education Department issued guidelines that outlined the way in which collection work should be undertaken. For the gathering of all fruits and plants, labour was to be mobilized in similar ways. This was most clearly described for the collection of brambles. Groups of pickers could be arranged to gather on "organized excursions" or "berry-picking rambles". Their work could form part of the school curriculum, as part of the nature study course, or as a recreational activity, during evenings, weekends, or holiday periods.[30] Not only were children joined by their teachers and youth organizers, but they were also accompanied by their parents. All gatherers were instructed to protect the environment, and were to safeguard agricultural land, crops and harvested foodstuffs. Because of the role which agricultural land played in the national war effort, they were instructed not to trespass on it, and were to obtain permission from "any farmers or other occupiers over whose land it is proposed to organize collection expeditions";[31] they were also forbidden to build fires or leave gates open.

The guidelines also tried to ensure that both fruits and plants were collected under optimum conditions. Memoranda outlined the time of year when work was to be undertaken, especially for the medicinal plants which had ingredients that were active only at certain times. During the early years of the Medicinal Plant Scheme, when a wide range of plant matter was gathered, collection took place throughout much of the year. Foxglove leaves were to be picked from "the middle of June to the end of July" when the plants were flowering; for rosehips that period was from "late August to the end of October"; nettles were to be picked "during the summer"; dandelion roots were to be dug "roughly from August to March". Only one plant, sphagnum moss, could be gathered "at any time".[32] For fruit, collection depended on ripeness. Rosehips, for example, were to be picked "when they are beginning to change colour until they are fully red"[33] or when "red and firm".[34] All fruit was to be harvested before it became overripe.

Fruit and plant matter was only to be obtained under certain conditions and handled in specific ways. All fruit was to be gathered during favourable weather, as "fruit picked immediately after rain will go mouldy and be useless for preserving". For those plants utilized for medicinal purposes, the wide range of plant matter and plant parts were gathered using a range of techniques.[35] Rosehips were to be collected free from bruising; no leaves and stems were to be included in the sample. Foxglove leaves could be "picked on a fine day any time after the dew has dried, and delivered at once for drying the same day"; they were not to be mixed with any other "kind of leaf". For gathering nettle leaves, pickers were instructed to wear "stout gloves" and use gardener's shears or a strong knife: "the plants should be tied in bundles of about 10–15 pounds and, after delivery, stood together in groups under cover to dry like sheaves of corn". Just as important as the need to pick the plant matter correctly, was that of delivering it properly to the processing centres. For the medicinal plants, this work ensured that the "principles" were not "destroyed" and the matter become "worthless commercially". Rosehips, for example, were weighed, placed into special hessian sacks, which held 28 or 56 pounds, and distributed at frequent intervals to allow the fruit to retain its levels of vitamin C.

The instructions offered by the central and urban directives for the collection, disposal and processing of the fruits and plants were critical to the success of the collection work and the general performance of the schemes. After the first year of their operation, when it was seen how arrangements had worked, the Scottish Education Department advised that their guidelines should be closely followed. The Department indicated that difficulties had occurred, especially in the work of gathering the medicinal plants. These difficulties had primarily resulted from problems in the relations between the collecting and processing areas. The Department argued for the need for close co-operation so that manpower and resources were not wasted.[36]

Outcomes of the partnerships

The partnerships established under the three schemes had a number of outcomes that benefited rural and urban areas in a number of ways. Owing to the different methods of handling and preserving the collected produce, that impact was not the same for each scheme. The most obvious impact of the partnerships between the urban directives and local networks was the successful gathering of plants and fruits. Geographically, these were collected over a wide area of Scotland. Indeed, the Scottish Education Department reported that the work was a "national activity".[37] Generally, they were secured from counties and districts that were predominantly rural, where agriculture formed the major industry. During 1946, the agricultural counties of Perth and Kinross secured 18.5 tons. Some rural parishes, with very small populations, such as Bolton, in East Lothian, gathered large totals. The 22 pupils at that school secured 500 pounds during the 1955 season, a bumper one.[38] However, some industrialized counties did make a significant contribution. Lanarkshire, for example, secured 24.5 tons in 1946, the largest single amount gathered in any county during that year.[39]

The collection work also supplied large quantities of plants and fruits. These amounts were often only expressed in general terms. In 1943, school children gathered a "considerable quantity" of plants for medicinal use[40] and in the two following seasons, their work resulted in "very gratifying results" being obtained.[41] Their response to the calls to gather plant material was also overwhelming at times. During the 1941 season, the collection of rosehips was "so successful" that the processing firm, Messrs Duncan, Flockhart and Company of Edinburgh "had to broadcast an appeal for the supply to stop".[42]

No detailed statistical data are available to show the full extent of the plants and fruits that were collected. Especially where gathered, processed and consumed on a local basis, the evidence is very fragmentary indeed. There is little evidence of the amounts of brambles and other wild fruits gathered and processed during any one year or throughout the operation of the schemes. Reports of statistics tend to cover a small geographical area, reflecting the nature of the work. However, there is a greater body of evidence for the medicinal plants, especially rosehips. So extensive was the total annual amount of fruit which was gathered, that it was expressed in terms of tons. Nationally, in each of the years between 1943 and 1945, between 70 and 77 tons were secured.[43] Cumulatively, that total was extensive, as was indicated by the *Glasgow Herald* in early 1947 when it reported that "during the past six years over 500 tons of rose hips have been gathered in Scotland".[44] However, collection levels were not at their highest during the war years, but afterwards, when the national food situation remained at a critical level. Not only was the largest single total, some 144 tons, recorded in 1947,[45] but figures remained significant into the 1950s, although they fluctuated greatly, from 46 tons in 1950, to 98 tons in the following year.[46]

The wild fruits and plants had an increasingly important role, both nationally and locally, during the war and afterwards until the food and economic situation became less difficult. At a national level, rosehips, as a welfare food, had a progressively centralized role within the work of the Medicinal Plant Scheme. The importance of its vitamin content ensured that it had an increasing number of roles in maintaining welfare. Hips took the place of vitamin rich foods that were no longer available in sufficient quantities, or had become absent altogether. In 1942, rosehip syrup was to "take the place of oranges and other juices for children"[47] so that they could be protected "against disease this winter".[48] By 1944 it was "being supplied to children as a substitute for fresh fruit".[49] Its importance was also seen in the extent of the annual rosehip collection, especially at the end of the war when the Scottish Education Department pointed out that there was a continued need to maintain supplies of rosehips and rosehip syrup. Not only was there a "continued urgent need for augmenting supplies of vitamin C" from that fruit[50] but also that "rose hip syrup remains one of the most important sources of vitamin C".[51] If rosehips had a growing significance, so too did the wild fruits. Their role was seen in the relationship that they had with supplies of cultivated fruits. The *Scottish Farmer* reported that cultivated fruits became available in even more limited amounts. In 1943, when fruit was "scarce",[52] the waste of fruit "would be inexcusable".[53] In the following year it noted further difficulties, as it was "more important than ever this year that no fruit should be wasted". But that position was even more problematic during 1945 when fruit was "scarcer than ever".[54] Increasingly, its household column advised women on the collection and use of wild fruits.

The work of the schemes benefited both rural and urban areas. As the majority of the collection work was undertaken in rural areas, that activity benefited them in a number of ways. Incentives were given for people to participate in the work, especially the gathering activities, of all three schemes. Gatherers were remunerated at a set rate of payment. A memorandum from the Scottish Education Department suggests that under the Medicinal Plant Scheme, processing firms paid set rates for material. One such price reported for rosehips stood at 2s. a stone (14 pounds) during the first years of collection.[55] Conversely, for wild fruit remuneration was given only when it was sold. Payment was made at a set price laid down in the annually revised government regulations for maximum soft fruit prices. Rates were set for a number of fruits, such as for crab apples and brambles, which stood at 2d. per pound in 1941; a maximum of 3d. per pound was set for fruit in general. For other wild fruits, rates were to be "agreed between the parties" who were selling and buying.[56]

The remuneration assisted and benefited the collectors. It had an impact on their personal and domestic economies and also on their local communities and the national economy. The "rosehip money" given to children for gathering rosehips was put to a number of uses.[57] School headmasters, who received the collected monies for their schools, decided upon their usage. At some schools, children were given the proceedings and were thus remunerated for their efforts. But schools also retained that money for a variety of purposes such as the "school fund account"[58] or its Red Cross fund, a national charitable cause.[59]

The consumption of fruits and plants, either unprocessed or processed, made a number of important contributions to food consumption patterns. Some was retained by the people who had gathered it. In such a case, it was incorporated within their domestic food supply, where it was eaten raw or processed into preserves, bottled, or used it for other purposes. In such cases, it largely benefited rural areas where the majority of the collection work took place. Outwith the domestic sphere, the fruit was sold through a number of channels that allowed it to be distributed throughout Scotland, and thus form part of the wider food supply. As a raw product, it was sold to both the wholesale or retail trades "for the benefit of the grower",[60] or to jam manufacturers located in large urban areas or close to them.[61] Fruit was sent to processing centres, which could, in turn, sell it.[62] When processed, preserves were also purchased by the Ministry of Food which pooled all supplies, and in turn, sold them at a fixed price.[63]

Within these channels of use, wild fruit (and also cultivated fruit) was incorporated into the food supply of a number of institutions, namely schools, hospitals, orphanages, communal feeding centres and military camps. Schools played an important role in processing and preserving the wild fruit as well as utilizing it by incorporating it into their food supplies. The gathering of fruit by children for their schools in rural areas is reported in school log books, such as that at Spott in East Lothian. There, on 18 September 1942, "the senior girls have made a collection of brambles which is being made into jam for their winter use".[64] But the impact of that fruit was far greater than this, as pooled fruit and preserves were also distributed over a wider geographical area. In East Lothian during 1943, the Education Sub-Committee recommended that fruit should be centrally collected at the school meals depot in East Linton, where it was to be manufactured into preserves to be distributed to schools.[65]

The preserves which were manufactured in schools and other institutions held a special status which allowed them to benefit their pupils and inmates. That resulted from the role which both sugar and preserves (a heavy consumer of sugar) played in the food supply and the national economy. Although sugar was rationed, schools and other preserving centres could obtain special allowances for preserving fruit.[66] These were set at maximum rates, normally a stated figure for each person who had fruit to preserve, and also a ratio of sugar to fruit.[6] Rationed amounts altered according to the availability of sugar, which changed seasonally and annually.[68] In addition, schools and other institutions also had further concessions over the use of their manufactured preserves, even though jam was rationed. As not all the preserves manufactured at the preserving centres were of a high standard, and were thus difficult to dispose of, schools and other institutions, such as orphanages and hospitals for children, were regarded as important consumption centres that could utilize the supply. In 1941 such supplies were not counted as part of "their allowance of jam under the Preserves (Rationing) Order".[69] These institutions were also given one further concession. Schools that manufactured preserves could add them to their ration. From 1942 onwards, they could be "retained for consumption by the school or school feeding centre over and above the authorized allowance of rationed preserves".[70]

The wild plants that were processed for medical uses benefited both rural and urban areas alike. Their impact was seen in terms of national rather than local impact and affected all areas of the country where the processed products were distributed. Rosehips that were manufactured into syrup by a central processor were distributed from a central location in Edinburgh to consumers throughout the country. When government departments and the newspaper presses reported the success of the rosehip collections each year, they did so in terms of national impact. That work ensured that rosehip syrup was widely available and that large supplies of it were manufactured as a welfare food.

The importance which the rosehip syrup had for the health of the Scottish population was widely acknowledged. The hips allowed a welfare food to be manufactured which was not only rich in vitamin C, but also had a far greater concentration than the standard substance for that source, oranges.[71] That figure was reported to be twenty times greater. Moreover, the syrup made an important contribution to the lives of vulnerable groups that required additional vitamins to maintain their health. The *Glasgow Herald* asserted how "the syrup made from these berries has done much to maintain the health of children and invalids during the past four years".[72]

Other plant matter collected under the Medicinal Plant Scheme had a national rather than local impact. The plant material secured an additional supply of drugs and medical provisions such as digitalis, from foxgloves, or dressings made of sphagnum moss, which helped to fill their increased demand owing to war-time conditions. But these plants also contributed to other aspects of British war-time policy. They assisted in saving shipping space and reducing imports, important aspects of economic policy which sought to make the country as self-sufficient as possible. In 1942, the Scottish Education Department suggested that some plants played an especially important role since "in normal times many hundreds of tons of shipping space were occupied by raw materials which could have been gathered in this country for the manufacture of drugs. Even stinging nettles were imported in large quantities."[73] Indeed, the nettle was one of the plants collected during the early years of the operation of the Medicinal Plant Scheme so that collecting plants saved foreign currency.[74]

Conclusion

This chapter has examined the relationship between rural and urban areas as production, processing and consumption centres during the Second World War and the immediately following years, through a case study of three government department organized schemes, which gathered and processed a range of wild fruits and plants: the collection of brambles, the collection of wild plants for medicinal purposes, and the processing of a range of surplus cultivated and wild fruits. The work of the government departments, especially the Scottish Education Department, involved the establishment of a range of partnerships at national, county, and local levels which linked rural and urban areas, resources and processors, supply areas and consumers. These relationships were not the same for all three schemes: they were shaped by the type of plant and fruit matter that was to be collected, methods of handling, and methods of processing (whether locally or centrally by specialist processors) and disposal into the food resources and utilization into the diet.

These partnerships were crucial to the success of the schemes. They widely publicized the work at national, regional and local levels. They allowed central directives, originating from urban areas, to be issued to rural areas where most of the wild fruits and plants were to be collected. These urban directives had a multifold role: they directed and ensured the plants were gathered in optimum conditions and at a time of the year when the active ingredients were at their strongest. They ensured that the fruits and plants were obtained using appropriate techniques, were delivered to local or central processing centres in the best possible condition and made arrangements for the processing to take place.

As a result of this collaboration, collection work, mainly undertaken in rural counties and their districts, was widely carried out and a large quantity of material was secured for foodstuffs, welfare foods and medical supplies.

These had specific roles to play across geographical areas of Scotland. Where processing took place at a local level, foods could be retained within a community; they could also pass into the general food supply within a district or county, and also nationally. By comparison, central processing ensured that distribution took place at a national level, and that both rural and urban areas benefited from the manufacturing work. In essence, the collaboration formed as a result of a difficult food situation during the Second World War and in the years immediately following, allowed wild plants and fruits to have growing importance, not only in rural areas but also in urban areas, and, moreover, nationally, throughout the country.

NOTES

1. NAS, SED, *Minute paper* (1940).
2. See for example McNeill (1929).
3. Beith (1995).
4. SED, *Memorandum* 229.
5. SED, *Circular* 503.
6. SED, *Memo.* 229.
7. See also SED, *Memo.* 20/1950.
8. SED, *Circ.* 493.
9. Scottish Home and Health Department (30 September 1953).
10. SED, *Memos* 204; 48/1943; 27/1945.
11. SED, *Memo.* 28/1952; Bolton Public School (14 November 1958).
12. SED, *Memos* 279A; 217.
13. Department of Agriculture for Scotland (1949).
14. Schools are noted from 1941 onwards. See for example SED, *Memo.* 217.
15. SED, *Memo.* 20/1950.
16. SED, *Memos* 204; 217.
17. SED, *Memo.* 204.
18. SED, *Memos* 48/1943; 37/1944; 36/1945.
19. SED, *Memo* 28/1950.
20. SED, *Memos* 21/1951; 28/1952.
21. Holmes (1999).
22. SED, *Report* (1941), 17; *Report* (1942); *Memo.* 42/1946.
23. SED, *Future of Women's Voluntary Services*, n.d. (c. 1945).
24. SED, *Memo.* 229.
25. SED, *Memo.* 42/1946.
26. SED, *Memo.* 21/1951.
27. Spott Public School (23 September 1941).
28. Spott Public School (9 July 1942).
29. Department of Agriculture for Scotland (1 May 1943).
30. SED, *Memo.* 229.
31. SED, *Memo.* 48/1943.
32. SED, *Memo.* 204.
33. *Glasgow Herald* (18 August 1942).
34. SED, *Memo.* 229.
35. SED, *Memos* 229; 204.
36. SED, *Memo.* 204.
37. SED, *Report* (1942), 12.
38. Spott Public School (7 October 1955).

39. *Glasgow Herald* (8 January 1946).
40. SED, *Memo*. 48/1943.
41. SED, *Memos* 37/1944; 36/1945.
42. *Glasgow Herald* (15 January 1942).
43. *Glasgow Herald* (30 August 1944).
44. *Glasgow Herald* (25 November 1947).
45. *Glasgow Herald* (8 January 1946); *Glasgow Herald* (25 November 1947).
46. SED, *Report* (1952), 29; *Memo*. 21/1951.
47. *Glasgow Herald* (5 September 1942).
48. *Glasgow Herald* (14 October 1942).
49. *Glasgow Herald* (30 August 1944).
50. SED, *Memo*. 42/1946.
51. SED, Memo. 45/1947.
52. *The Scottish Farmer* (11 September 1943).
53. *The Scottish Farmer* (8 May 1943).
54. *The Scottish Farmer* (7 July 1945).
55. SED, *Memo*. 204; SED, *Memo*. 229; *Glasgow Herald* (24 September 1941); *Glasgow Herald* (5 September 1942).
56. SED, *Memo*. 229.
57. Bolton Public School (14 December 1954).
58. Bridge of Weir Public School (7 November 1941).
59. Barrhead-Grahamston Public School (3 November 1941).
60. *The Scottish Farmer* (13 July 1940).
61. *The Scottish Farmer* (2 September 1944).
62. SED, *Memo*. 217.
63. *The Scottish Farmer* (13 July 1940).
64. Bolton Public School (18 September 1942).
65. East Lothian Education Committee (10 June 1943).
66. Department of Agriculture for Scotland (1 June 1940).
67. SED, *Memo*. 50/1943.
68. Murray (1975), 770–776.
69. SED, *Memo*. 217.
70. SED, *Memos* 279; 50/1943.
71. *Glasgow Herald* (24 September 1941).
72. *Glasgow Herald* (8 January 1946).
73. SED, *Memo*. (13 May 1942).
74. SED, *Memo*. 204.

BIBLIOGRAPHY

Abbreviations:
GCA: Glasgow City Archives
NAS: National Archives of Scotland

Barrhead-Grahamston Public School, *School Log Book* (3 November 1941) GCA, CO2/5/6/4/4.

Beith, Mary, *Healing Threads: Traditional Medicines of the Highlands and Islands* (Edinburgh, 1995).

Bolton Public School, *School Log Book* (18 September 1942, 14 December 1954, 14 November 1958) NAS, CO2/5/4/1.

Bridge of Weir Public School, *School Log Book* (7 November 1941) GCA, CO2/5/6/10/2.

Department of Agriculture for Scotland, *Agriculture in Scotland. The Report of the Department of Agriculture for Scotland for the Ten Years 1939–1948* (Edinburgh, 1949).

Department of Agriculture of Scotland, *Marketing of Fruit and Vegetables: Emergency Organization*, To the Secretary, no. 64 (1 June 1940) NAS, AF47/7/6/1.
Department of Agriculture of Scotland, *Marketing of Fruit and Vegetables in War Time*, Ministry of Food Press Notice (1 May 1943) NAS, AF47/7/6/4.
East Lothian Education Committee, *East Lothian Education Committee. Minute Book* (18 September 1942) NAS, CO7/5/1/18.
East Lothian Education Committee, *East Lothian Education Committee. Minute Book* (10 June 1943) NAS, CO7/5/1/17.
Glasgow Herald 1941–1947.
Holmes, Mrs A, Pilmuir Farm, Balerno, Midlothian, oral evidence, collected 1999.
McNeill, Marian, F, *The Scottish Kitchen. Its Traditions and Lore with Old-Time Recipes.* (Edinburgh, 1929).
Murray, Keith, A. H., *History of the Second World War: Agriculture* (London, 1975).
Scottish Education Department (SED)
Circular 493 (31 August 1917) NAS, ED44/1/8.
Circular 503 (9 August 1918) NAS, ED44/1/8.
Education in Scotland. A Report of the Secretary of State for Scotland (Edinburgh, 1952).
Food Production in Schools. Minute paper to Mr Corsfield (5 July 1940) NAS, ED48/180/3.
Future of Women's Voluntary Services, n.d. (c. 1945) NAS, ED24/35.
Memorandum 204 (4 June 1941) NAS, ED44/1/12.
Memorandum 217 (22 July 1941) NAS, ED44/1/12.
Memorandum 229 (1 October 1941) NAS, ED44/1/12.
Memorandum (13 June 1942) NAS, ED44/1/13.
Memorandum 279 (7 August 1942) NAS, ED44/1/13.
Memorandum 279A (11 September 1942) NAS, ED44/2/2.
Memorandum 48/1943 (26 May 1943) NAS, ED44/1/14.
Memorandum 50/1943 (31 May 1943) NAS, ED44/1/14.
Memorandum 37/1944 (25 April 1944) NAS, ED44/1/15.
Memorandum 27/1945 (16 April 1945) NAS, ED44/1/16.
Memorandum 36/1945 (16 May 1945) NAS, ED44/1/16.
Memorandum 42/1946 (17 June 1946) NAS, ED44/1/17.
Memorandum 45/1947 (23 June 1947) NAS, ED44/1/18.
Memorandum 20/1950 (30 May 1950) NAS, ED44/1/21.
Memorandum 28/1950 (21 July 1950) NAS, ED44/1/21.
Memorandum 21/1951 (28 June 1951) NAS, ED44/2/11.
Memorandum 28/1952 (4 July 1952) NAS, ED44/2/12.
Summary Report on Education for the Years 1939 and 1940 (Edinburgh, 1941).
Summary Report on Education in Scotland for the Year 1941 (Edinburgh, 1942).
The Scottish Farmer 1940–1945.
Scottish Home and Health Department, *Future of the Welfare Foods Service* (30 September 1953) NAS, HH64/193.
Spott Public School, *School Log Book* (23 September 1941) NAS, CO7/5/4/3.
Spott Public School, *School Log Book* (7 October 1955) NAS, CO7/5/4/4.

MARJATTA HIETALA

14 Food and natural produce from Finland's forests during the twentieth century

The Nordic natural environment of cold climate, long winters, lakes and rivers freezing over, alternation of darkness and light are the elements, which have shaped the Finnish landscape for centuries. Climate and nature are key factors in analyzing the interaction between physical surroundings and food culture in Finland. The growing season is very short, there being only 150 or 200 days between spring and autumn ploughing.[1] In Southern Finland, the average temperature is between 12 and 22 degrees Celsius in July and in January between –9 and –4 degrees.

Finland[2] is a very green country: of the total area, 68 per cent is covered by forest, 10 per cent is water and six per cent cultivated land. For centuries Finns have lived off the forest and its resources: during earlier centuries from hunting and during the last century from wood, pulp and the paper industry. After the Second World War Finland industrialized rapidly, which was followed by heavy migration into cities, especially in the 1960s. However, the highly urbanized population is still able to experience the surrounding wilderness and forests. It is estimated that an average Finn makes 20 visits to forests each year, excluding those for picking fruit and hunting activities.[3]

Today, forests have multiple uses in Finland of which wood production, comprising, for example, sawn timber, pulp and paper, as well as forest-based energy production, is superior to any other single use. In the mid-1990s a working group estimated that the Gross National Product of all activities related to the gathering, trade and processing of these products was as much as 1.8 billion FIM[4] and 0.4 per cent of the GNP of Finland. Derived from that figure a total employment estimate of 900 persons in export-related activities and a total of 9,000 person-years (including the collection for household use) was given.[5] Forests are used for carbon sequestration, nature-based tourism, reindeer husbandry, biodiversity conservation and outdoor recreation which all seem to have considerable economic importance. Additionally, there are at least one million summer houses or cottages in Finland. Finns are also looking into sustainable ways to manage forests so that they will be well preserved for future generations.

Moreover, forests and nature have had, and still have, an aesthetic function in the Finnish culture. Both Art Nouveau at the turn of the twentieth century and post-war Modernism, especially Finnish Design, drew its motifs from nature. Authenticity and closeness to nature were prominent as distinctive Finnish qualities together with simplicity and modernity. These qualities were used in the construction of Finnish identity also.[6] In the musical field, Jean Sibelius, the best-known Finnish composer, looked to nature for inspiration in his creation of a national musical tradition in Finland, for example, in *Finlandia* (1899).[7]

This chapter analyzes the role products from the wild, especially berries and mushrooms, have had in Finnish everyday life from consumption and export to mythology. It explores how much picking berries and mushrooms has been a resource for living as well as a part of Finnish diet. Production and trade of nature-based products are shown by statistics, but it is difficult to estimate the consumption of berries and mushrooms in individual households.

Everyman's Rights in Finland

Everyman's Rights is a special Nordic 'law', which gives everybody the right of public access to the nature. The origin of the 'law' is to be found in the Middle Ages when there were great uninhabited forests around the Nordic villages. Until the beginning of the twentieth century forests were seen as public property, although a distribution of forest land was made in the general parcelling out of land in the eighteenth century. Still, taking wood from forests to make tools and household utensils was allowed, based on Everyman's Rights.[8] Nowadays, in forests everybody is allowed to pick fallen branches and cones; and, naturally, to pick berries and mushrooms regardless of whether the land is owned by private persons, industrial companies or the state.[9]

The special character of the custom of Everyman's Rights is that it is not a part of the Finnish legislation as such. Some of it is included in legislation, such as the right to wander in the wilderness, which dates as far back as 1734,[10] or that it involves everybody visiting Finland, regardless of his or her nationality. It is included in legislation that, according to Everyman's Rights, trespassing into people's gardens and yards is forbidden, though the exact limits are not mentioned. In part, Everyman's Rights is a common law of customs based on the national values and Finnish culture.[11] It gives lots of freedom to enjoy and use natural resources but, since nature is seen as common heritage, it entails also lots of responsibilities. For example, wandering has to be done in such a way that it does not harm the natural environment in any way. Thus, it is stipulated in the legislation that moving on somebody else's ground that is in its natural state is allowed on foot, on skiis or by riding a horse or a bicycle. Also, people are expected by the law to leave no litter in the woods and otherwise take care that no trace of their visit is left behind.[12]

Everyman's Rights and its definition have raised some controversies, especially in Letters to the Editor in Finnish newspapers. In the letters' columns there is a continuing discussion about the extent from people's courtyard areas that Everyman's Rights covers. Also, people express their worries about the future of public property and national heritage. It is feared that a multitude of foreigners enjoying Everyman's Rights will put some strain to Finnish natural resources. It is argued that Everyman's Rights is not learned from books but absorbed as part of the acculturation process for Finns. Thus, the responsibilities entailed in Everyman's Rights might be something that foreigners will not be aware of. This question was raised particularly when Finland joined the European Union, which was expected to increase the number of foreigners and tourists visiting Finland.

Importance and consumption of wild berries and mushrooms in Finnish diet

There are today 50 different wild berries that grow in Finland, 37 of which are edible. Furthermore, there are 200 edible mushrooms of which 22 have been categorized as commercial mushrooms. It is estimated that the total crop of edible mushrooms varies from 350,000 tons to 1,000,000 tons per annum and wild berries approximately between 500,000 and 1,000,000 tons yearly. Still, only less than one percent, approximately 2,000–10,000 tons, of the total crop of edible mushrooms is picked on average every year. Of the most frequently picked wild berries, lingonberries and blueberries, only less than one-tenth of the total crop is picked yearly; of the cloudberries the percentage is higher.[13] Finns consume approximately eight kilograms of wild berries per person per annum. The highest recorded amounts of picked wild berries are from 1997 and 1998 when the amounts reached 50,000–57,000 tons. An earlier record was made during and after the Continuation War in 1943–1947, of 47,000 tons.[14] In addition to berries and mushrooms, natural herbs are also picked and marketed in Finland. Precise trade statistics are not available for those quantities. Also lichen and moss are gathered.

In Finland, as elsewhere in Europe, food quality and preservation issues have been broadly investigated over the years; for example, the State Technical Research Centre examined these questions from a Finnish perspective in the 1950s and 1960s.[15] Studies of berries have focused in particular on the vitamin contents of and methods for preserving the flavour of frozen products. Traditionally, there have always been differences in the usage and preservation of wild berries and mushrooms between eastern and western Finland. Mushrooms and berries belonged more to the eastern tradition. Salted mushrooms were used as substitutes for meat during winter and spring. In eastern areas, also, fresh bread was baked every week instead of using dry bread in the diet. Even today, rural households in eastern Finland are better equipped with bigger freezers of 500–1,000 litres in capacity than western households. All in all, according to the Household Expenditure Survey of the Statistical Office of Finland in 1990, 80 per cent of all Finnish households owned freezers.[16] Today frozen berries and mushrooms are part of the staple diet of every Finn – purchased in urban market places, or even better, picked by oneself.

Currently, scientific research is interested in the nutritional values of berries, especially in the effects of micronutrients on people's health; however, the health effects of wild berries have been common knowledge among people for centuries. The newspaper and educational writings of Elias Lönnrot, a doctor of medicine and scholar of folklore, who collected the poems in the Finnish national epic, *Kalevala*, influenced people's food habits from the beginning of the nineteenth century onwards. In his book *Flora Fennica* Lönnrot introduced species found in Finland and gave advice for using wild berries and herbs in folk medicine and as food supplements.

In his writings, Lönnrot looked for means of survival in difficult times. Tree bark had been used for centuries to adulterate bread dough. Bark bread (*pettuleipä*) was baked especially during the years of crop failure and directly after them, for example in 1866–1868 when Finland had to face a severe famine. In those times, the dream of every Finnish peasant was to have bread that had been baked from pure flour; flour that had not been adulterated with "bark from a pine tree by a half". In addition to bark bread, Lönnrot gave advice on baking bread from roots, from seeds of grass or from lichen. Additionally, he wrote that potatoes and turnips could be used to adulterate bread dough, as could also ground berries, such as rowan berries, lingonberries and blueberries as well as dried and ground-up sphagnum and lichen. Lönnrot applied foreign know-how in his writings, for example, in Copenhagen, he noted, good bread had been baked from dough that comprised of one-third of rye straw chaff and two-thirds of rye flour. In close proximity to the city of Dijon, chaff had been ground and sifted to make beautiful flour that appeared similar to real rye flour. Furthermore, bean and pea stems could also be used to make flour for baking bread. Savoury bread was baked when flour was added to juice cooked from birch sap, bone stock or linseed. Furthermore, if potato porridge was added, the bread was even better.[17]

Lönnrot was familiar with the positive effects of berries on health, even when there was yet no scientific knowledge of vitamin C.[18] Of all the wild berries, Lönnrot had the highest opinion of lingonberries and cranberries. He said that lingonberries were excellent in quenching thirst and cooling oneself. Lingonberry water, which was easily made by pouring spring water on top of berries and adding little sugar, was widely used in curing fever. Lingonberries were used in making wine, in addition to jam, that was especially tasty and healthy. Lingonberry leaves were used to make tea. Lönnrot praised syrup made from cranberries that was also effective in curing fevers, as were dried blueberries whose positive effects Lönnrot found out in the 1860s. Savoury porridge could be made from blueberries by adding them to rye flour. Moreover, Lönnrot taught that mixing blueberries, lingonberries and rye flour could make good sour bread.[19] Interestingly, porridges made from rye flour, lingonberries or blueberries are still part of the Finnish diet as well as blueberry pies that were traditionally baked in Finnish homes on Saturdays. It is worth noting also that red, white and black currants grown in gardens have supplemented, but not displaced, wild berries in Finland.

In 1936 the Finnish Government appointed a special Committee for Nutrition of the Population. In 1936–1937 the Committee ordered a study of the diet of the low-income rural population, which concluded that 51 per cent of the studied families (288 persons) received less energy than they needed.[20] At the time the food intake of Finns consisted of grain products (43 per cent), dairy products (26.5 per cent), potatoes (9.5 per cent), meat (9 per cent), sugar (6 per cent), eggs (0.2 per cent), and finally, vegetables, fruit and berries (2 per cent) (see Table 14.1).[21]

Table 14.1 *The percentage share of different foodstuffs from the total energy intake in the Finnish diet in 1936–1937 and 1950–1951.*

Foodstuff	Committee 1936–1937 (%)	Roine 1950–1951(%)
Grain and grain products	43.0	34.0
Potato	9.6	6.5
Other vegetables	1.3	1.0
Fruit and berries	0.6	1.5
Sugar	6.1	10.5
Margarine	2.4	3.0
Butter	7.5	13.4
Other milk products	19.2	20.0
Meat and meat products	9.3	7.5
Fish	0.9	1.2
Eggs	0.2	1.4
Total	**100.0**	**100.0**

Source: Virtanen & Turpeinen (1940); Roine (1954), cit. in Hietala (1995).

In 1954 Professor Paavo Roine published his study of the Finnish diet in 1950–1951, which covered 180 families from different parts of the country. At the time 54 per cent of the total energy intake of Finns came from carbohydrates, 34 per cent from fats and 12 per cent from protein. The research also indicated that between 1936–1937 and 1950–1951 the percentage share of grain products in the total energy intake had diminished from 43 to 34, whereas the share of dairy products increased from 26.7 to 33.4 (see Table 14.1). Moreover, the share of sugar had grown from 6.1 to 10.5 per cent and the consumption of eggs from a mere 0.2 to 1.4. Meat, on the other hand, comprised only 7.5 per cent of the total energy intake in 1950–1951 compared to 9.3 per cent in 1936–1937. In addition, it was concluded that an alarming share of the total energy intake came from fats. Nevertheless, the consumption of vegetables, fruit and berries had increased by half a per cent to 2.5. If fruit and berries are examined separately from vegetables, they comprised 0.6 per cent of the total energy intake in 1936–1937 only but 15 years later already 1.5 per cent. The most common fruit consumed in 1950–1951 was apples (14 g per person per day) followed by lingonberries (11 g), and garden berries (10 g).[22]

In 1951, Finnish Nobel Prize winner A. I. Virtanen,[23] together with Paavo Roine, introduced the so-called food circle (*ravintoympyrä*) with six segments that were: 1. milk, cheese; 2. fats; 3. grain products; 4. potatoes; 5. meat, fish, eggs; 6. root crops, vegetables, berries, fruit (see Figure 14.1).[24] The low share of vegetables, fruit and berries in the Finnish diet concerned Virtanen and Roine: "By adding vegetables, fruit and berries into our diet, there would be more variety in tastes [...] and more vegetables, fruit and berries would bring lightness into our otherwise rather heavy diet. Moreover, they would ensure the necessary intake of vitamin C and B vitamins as well as minerals".[25]

Figure 14.1 *The food circle with six segments.*

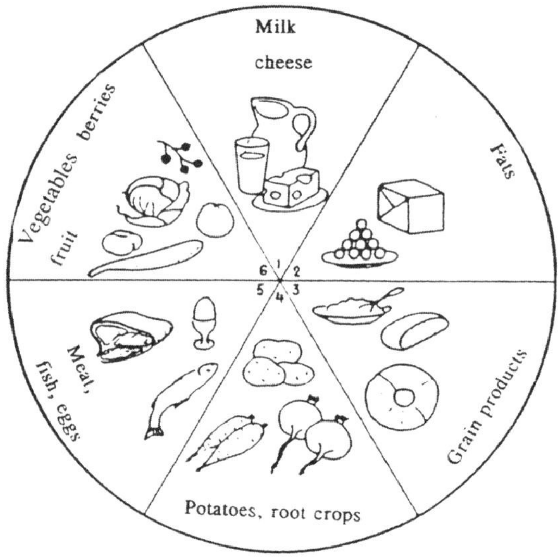

Source: Virtanen & Roine (1956).

The introduction of the food circle was one example of the attempts by health authorities to influence people's nutritional habits with efficient counselling and educational material. Similar attempts were the Project on the Nutrient Intake of Finns in 1967–1969, funded by the National Nutrition Board, and its follow-up survey, conducted in 1973–1976, in co-operation with the Finnish National Pensions Institute. The surveys indicated that since the beginning of the twentieth century the consumption of grain products and potatoes had fallen by half. The consumption of fats, meat products, vegetables, eggs and sugar, on the other hand, had steadily grown; meat and meat products especially were consumed heavily in the 1960s, which was in connexion with increasing standard of living.[26]

The Central Statistical Office of Finland has estimated the consumption and expenditure of Finns in the Household Expenditure Surveys it has conducted since the 1970s. In addition to purchased commodities, the value of own produce, or that received in kind – agricultural, garden and gathered products – were included in evaluating consumption (valued at the producer price). In 1971 foodstuffs formed a quarter of the total household expenditure; berries and fruit constituted then 3 per cent of the household expenditure, growing by 0.3 per cent during the next five years.[27]

Commercial use of nature-based products

Income from picking berries and mushrooms sold through organized market channels has varied annually between 30 and 142 million FIM in 1977–1995. Commercial picking of berries has been high until the recent years although the real prices paid to the pickers have decreased. Picking wild berries and mushrooms is tax-free income, which has been important for families of low income and for families with many children. The majority of wild berry and mushroom pickers are in the ages between 25 and 65; women do most of the picking. The majority of this income goes to the sparsely populated northern and eastern parts of the country (see Table 14.2). The most ardent 'hunters' of mushrooms are 55 to 66 year-olds in eastern, central and northern Finland. Moreover, people living in those parts of Finland also consume more mushrooms than other Finns. Annual household consumption of mushrooms varies between one-half and two kilograms per person. Moreover, the purchasing and trading of natural produce has been concentrated in these areas. In 1997, for instance, there were a hundred times more flows of wild blueberries from Lapland, and Oulu and Kainuu regions than from the West. It has been estimated that currently there are 50–100 full-time entrepreneurs in the mushroom business. Further industrial processing is concentrated in southern Finland; and the most significant frozen-mushroom businesses are joint Italian-Finnish companies.[28]

Table 14.2 *Flow of wild berries and mushrooms to market, by region and in the whole country in 1997 (in 1,000 kg).*

Area	Wild berries				Mushrooms			
	Blue-berry (1,000 kg)	Lingon-berry (1,000 kg)	Cloud-berry (1,000 kg)	Total (1,000 kg)	Lact. Agarics (1,000 kg)	Boletus (1,000 kg)	Chantarelle (1,000 kg)	Total (1,000 kg)
Western Finland	29.5	820.1	4.6	854.2	31.4	11.6	1.6	44.6
Eastern Finland	397.9	916.7	8.0	1,322.6	320.5	174.8	0.9	496.2
Oulu and Kainuu Region	2,058.5	4,665.7	71.1	6,795.3	3.1	2.3	0.0	5.4
Lapland	477.2	1,210.9	243.1	1,931.2	0.3	0.3	0.0	0.6
Whole country	2,963.1	7,613.4	326.8	10,903.3	355.2	189.1	2.5	546.8

Source: Food and Farm Facts Ltd 22.3.1999.

Table 14.3 *Income from picking wild berries, forest mushrooms and garden berries in 1977–1998.*

Year	Wild berries (1,000FIM)	% of total	Forest mushrooms (1,000FIM)	% of total	Garden berries (1,000FIM)	% of total	Total (1,000FIM)
1977	72,295	87	719	1	10,233	12	83,247
1978	45,805	76	2,593	4	12,313	20	60,771
1979	50,071	72	1,720	3	17,332	25	69,123
1980	57,755	76	1,915	3	15,983	21	75,653
1981	35,277	51	3,955	6	29,391	43	68,623
1982	53,235	50	2,454	2	51,711	48	107,400
1983	89,447	60	3,811	3	55,163	37	148,421
1984	31,601	36	3,340	4	53,813	60	88,754
1985	54,907	42	8,588	7	65,236	51	128,731
1986	44,669	41	4,396	4	60,058	55	109,123
1987	41,982	47	6,504	7	40,606	46	89,092
1988	85,304	62	11,646	8	41,140	30	138,090
1989	56,786	51	3,727	3	50,357	46	110,870
1990	52,073	48	6,801	6	48,965	46	107,839
1991	73,554	60	10,266	8	39,462	32	123,282
1992	45,667	49	6,001	7	41,242	44	92,909
1993	27,220	39	3,086	4	39,157	57	69,463
1994	136,276	80	5,919	4	26,780	16	168,975
1995	58,725	70	4,253	5	20,640	25	83,618
1996	52,158	63	4,090	5	25,960	32	82,209
1997	62,708	65	7,003	7	24,662	26	94,373
1998	51,804	56	13,625	15	26,450	29	91,879

Source: Food and Farm Facts Ltd 22.3.1999.

Note: Wild berries: 1977–1985 included only blueberries, lingonberry and cloudberry, from 1986 also rowanberry, buckthorn, crowberry and cranberry and arctic bramble. Forest mushrooms: 1977–1985 included chantarelle, boletus, milk caps, from 1986 also false morel. From 1988 included also horn of plenty and other edible mush rooms. Garden berries: 1977–1985 included strawberry, black currant, red and white currant. From 1986 also gooseberry and rasberry and aronia.

Table 14.3 illustrates the importance of the income from picking wild berries when compared to that from garden berries. The percentage shares indicate that at the end of the 1970s considerably more income originated from picking wild berries when compared to the utilization of garden berries, whilst in the 1980s the numbers had been turned around (the most striking examples being the years 1984, 1985, 1986 and 1993). In part, the increased cultivation of strawberries explains this change. In the mid-1990s again, the percentage shares indicate a change: income from picking wild berries was especially high. One explanatory factor underlying this might be the economic recession Finland was undergoing at that time. Picking berries and mushrooms is understandably dependent on weather conditions; for example, the summer of 1998 was especially productive for mushrooms when there was rain and warmth in suitable proportions. Then, the income was almost double that of the previous year.

The Official Statistics of Finland show that the most exported berry has been the lingonberry. There are no specified statistical sources for the export of berries before the Second World War, nor for the time between the 1970s and 1980s. From the 1920s and 1930s we can find statistics of the export of fruits and berries measured together, which indicate that their export increased four-fold from 1921 to 1931. Table 14.4 shows that lingonberries were exported approximately 2,000 tons yearly during the 1950s and 1960s.

Table 14.4 *Nature-based exports: fresh lingonberries, cloudberries and lichen (1,000 kg).*

Year	Lingonberry	Cloudberry	Lichen
1951	1,862	40	1,208
1961	2,317	n.a.	2,322

Source: Statistical Yearbook of Finland (1952), 1962.

The export of natural products has mainly been a trade in raw materials. Table 14.5 demonstrates how the new freezing technology has changed the export of wild berries: currently, trade in blueberries is mostly carried out in frozen goods. Both export and import amounts in recent years have been affected by competition for raw material prices from Sweden, the Baltic countries, Russia, Poland and China. Additional competition that diminishes Finland's berry markets in Europe and Japan comes from North America, where 'wild' berries are cultivated. Of the fresh Finnish wild berries, only the lingonberry is exported more than imported, mostly to the Scandinavian countries, Germany and Estonia; recently though, the import of lingonberries has been growing faster than export. The same applies to fresh blueberries and cranberries.[29]

Table 14.5 *Import and export of wild berries in 1996–1998, amounts in 1,000 kg.*

Year	Fresh blueberries		Frozen blueberries		Fresh lingonberries	
	Import	Export	Import	Export	Import	Export
1996	259.3	199.9	567.1	1,532.5	41.8	394.0
1997	263.3	0.9	1,255.5	1,699.0	146.9	881.5
1998	258.5	139.4	1,288.6	1,431.8	550.2	1,236.7

Source: Food and Farm Facts Ltd 22.3.1999.

Berries and mushrooms in Finnish mythology

Nature-based products have a role in Finno-Ugric mythology. From Eugen Helimski's study[30] into shamanism we find that mushrooms in conjunction with alcohol were used to help to produce the mental state in which the spirits were able to 'sing' through the shaman; thus, there is a strong link in Finno-Ugric languages between the words for mushrooms and alcohol. For example, in Ngansan, the word for being intoxicated and drinking alcohol is etymologically directly linked to the word for mushrooms.

Another mythological story in which the nature-based products play a vital part is the 'berry legend', a story of the miracle of the Virgin Mary's impregnation by eating a berry. In the story, Mary eats a lingonberry – although interestingly it is translated into English as a cranberry – the berry goes down to her belly and grows there into a baby. The same story has been widely circulated but in the old Finnish folk poems it has an interesting Baltic-Finnish adaptation: the language and the story line, as well as the characters and their milieu, were affected by the local circumstances and culture in reiterations of the poem in Finland. For example, at the end of her pregnancy Mary is looking for a sauna that is the traditional place for childbirth in the Baltic-Finnish area. It is thought that the misunderstanding of Mary becoming pregnant by eating a berry was a result of local poet-singers' misconception of church paintings depicting the flight into Egypt. In Finnish versions of the story, the fruit Mary eats is an apple in southern Finland, whereas in northern Finland, where apples were not known, she eats a wild berry. This interpretation of the fruit into a berry introduced another dimension to the story in the Finnish language. The homonymic conformity of the word for a berry, *marja*, and a woman's name, *Marja*, appears to explain why in Finnish tradition the berry came to be seen as the source of impregnation. This conformity gives the impression that Jesus was born from a wild berry.[31] In symbolic language, the original fruit for impregnation has been an apple; that is found in central European and English ballads.[32]

Nature and Finnish identity

One objective in this chapter is to investigate whether picking berries and hunting for mushrooms is a survival from poor economic times or whether it belongs to Finnish identity as such. Fortunately, Finnish identity can be studied outside the borders of Finland. The 'Finnish identity' of the Finnish-Americans in the Minnesota area can be used as an example. In the booklet *Fantastically Finnish, Recipes and Traditions*, the traditional Finnish love for natural ingredients in the diet can be still found in the United States. "Finns have always appreciated basic, straightforward, natural flavours, textures and appearances in their food. Food is a gift, to be tampered with as little as possible, so that the God-given flavours still remain true, and the nutritional qualities stay high [...] Whole grains, exemplified in their breads and porridges, fresh vegetables, potatoes, wild berries, mushrooms, game and fish, all figure highly in the Finnish diet."[33] Everyman's Rights is not mentioned by name but referred to in the appreciation of the harvest in Finnish forests and gardens: "Summer is berry picking time. Arctic cloudberries, strawberries, blueberries and lingonberries flourish. Finns are avid mushroom hunters in spring and fall. Gardens yield fresh vegetables and potatoes. Fresh fish, caught in the Baltic, are sold and eaten raw, pickled or smoked. Wild game is hunted in forest-covered Finland."[34] These praises for the nature's gifts in Finnish diet – still evident among the third and fourth generation Finns – seem to suggest that the love for natural ingredients is not merely an outcome of the life under severe conditions, but something inherent in the Finnish identity. Closeness to forests and trees still shapes and moulds things Finnish, including people.

Future

Everyman's Rights are currently enjoyed by a large majority of Finns. On average Finns exercise, play and spend time out of doors three to four times a week all the year round, with cross-country skiing being the most popular winter-activity. Furthermore, over half the people spend their leisure time in 'far-away' summer cottages. Wandering in the forests is an experience in itself; they are places for meditation and relaxation. Nine out of ten outdoor activities in Finland takes place in recreation areas or natural tourist sites where there is forest.[35] Although forest-based products such as sawn timber, pulp and paper, and energy production will continue to be vital for the Finnish economy, at the same time the recreational use of forests and nature, for example picking berries and gathering mushrooms, is ever more important for urbanized Finns.

NOTES

1. Mead (1981).
2. The area of Finland is 338.145 km2 which makes it the fifth largest country in Europe, excluding the Russian federation. Roughly one-third of the country lies north of the Arctic Circle. Finland has only five and a half million people and the density of population is low (17 inhabitants per km2 and 205 inhabitants per km2 in Uusimaa, the southern part of the country).
3. Saastamoinen (1997).
4. 1 FIM equals approximately 0,168 €.
5. Saastamoinen (1997), 115–116.
6. Hietala (2002), 235.
7. Alho, Hawkins & Vallisaari (1997), 290.
8. Vilkuna (1979).
9. Palokangas, R., Tarukannel, V. & Nuuja, I., (1993); Saastamoinen (1997).
10. Laaksonen (1980).
11. Vuolle & Oittinen (1994).
12. Ibid.
13. See http://www.mmm.fi/luonnontuote/luonnontuotetietoa.htm, printed 4.8.2002.
14. Ibid.
15. Michelsen (1993), 51–58
16. Hietala & Lepistö (1998).
17. Lönnrot, cit. in Majamaa (1992b), 26–38.
18. See also Rautavaara (1980), 107.

19. Lönnrot, cit. in Majamaa (1992a), 91–92.
20. Hietala (1995), 156–167.
21. Virtanen, & Turpeinen (1940), 8–126.
22. Roine (1954), 73–74.
23. A. I Virtanen developed the production method of AIV-silage. This innovation improved the production structure of the Finnish farm industry and opened a route to the world market for products like butter and cheese. In the 1930s Professor Virtanen focused his research on the biochemistry of vitamins. Hietala (1995), 164.
24. Cf. ch. 4 in this volume.
25. Virtanen & Roine (1956), 86–87.
26. Koskinen (1974).
27. Central Statistical Office of Finland (Helsinki, 1974 & 1979).
28. See http://www.mmm.fi/luonnontuote/luonnontuotetietoa.htm, printed 4.8.2002.
29. See http://www.mmm.fi/luonnontuote/luonnontuotetietoa.htm, printed 4.8.2002.
30. Helimski (1998).
31. Kuusi (1977).
32. Sarmela (1994), 217.
33. Ojakangas (1985).
34. Ibid.
35. See *Finnish Statistical Yearbook of Forestry*, 201.

BIBLIOGRAPHY

Alho, Olli, Hawkins, Hildi & Vallisaari, Päivi (eds), *Finland: a cultural encyclopedia* (Helsinki, 1997).

Finnish Statistical Yearbook of Forestry, Official Statistics of Finland, Agriculture, forestry and fishery 14 (Helsinki, 2000).

Food and Farm Facts Ltd (22.3.1999).

Helimski, E., *Samojedit ja šamanismi. Viisi luentoa samojedeista, šamanismista ja uralilaisesta kulttuurista* [Samojeds and shamanism. Five lectures on Samojeds, shamanism and Ugric culture] (Tampere, 1998).

Hietala, Marjatta, 'From famine to welfare: food patterns in Finland during the past hundred years', in: Adel den Hartog (ed.), *Food Technology, Science and Marketing: European Diet in the twentieth century* (East Linton, 1995), 156–167.

Hietala, Marjatta, 'Nature and the city milieu', in: Marjatta Bell & Marjatta Hietala (eds), *Helsinki – The Innovative City – Historical perspectives* (Helsinki, 2002), 235–237.

Hietala, Marjatta & Lepistö, Vuokko, 'Arctic Finland and the new technology of food preservation and refrigeration, 1850–1990', in: M. R. Schärer & A. Fenton (eds), *Food and Material Culture* (East Linton, 1998).

Koskinen, Esko H., *Suomalaisten ruoankäyttö ja ravinnonsaanti vuosina 1967–1969* [Finnish diet and nutrient supply in 1967–1969] (Helsinki, 1974).

Laaksonen, K., *Toisen maan yleiskäytöstä* [Using Somebody Else's Land], The University of Helsinki, the Department of Private Legislation, Publication B 25 (Helsinki, 1980).

Kuusi, M., *Finnish folk poetry: epic* (Helsinki, 1977).

Lönnrot, Elias, *Flora Fennica*, in: Raija Majamaa (ed.), Elias Lönnrot, Valitut teokset 4: Ohjeita ja runoelmia [Selected works 4: writings and poems] (Helsinki, 1992a).

Lönnrot Elias, *Hyvänsuotuisia neuvoja katovuosina* [Advice for years of crop failure], in: Raija Majamaa (ed.), Elias Lönnrot, Valitut teokset 4: Ohjeita ja runoelmia [Selected works 4: writings and poems] (Helsinki, 1992b).

Mead, W., *A historical geography of Scandinavia* (London, 1981).

Michelsen, Karl-Erik, *Valtio, teknologia, tutkimus. VTT ja kansallisen tutkimusjärjestelmän kehitys* [State, technology and research. Technical Research Centre of Finland and the development of national research system] (Helsinki, 1993).

Ojakangas, B., 'The Food Traditions of Finland and Favored by Finns', in: J. Zug & S.

Roemig (eds), *Fantastically Finnish. Recipes and Traditions* (Iowa, 1985).
Palokangas, R., Tarukannel, V. & Nuuja, I., *Ympäristönsuojelun hallinto ja lainsäädäntö* [The Environmental Administration and Legislation] (Jyväskylä, 1993).
Rautavaara, Toivo, Miten luonto parantaa. *Kansanparannuskeinoja ja luontaistuotelääketiedettä* [Nature's cures. Folk and natural medicine] (Porvoo, 1980).
Roine, Paavo, 'Työläis- ja toimenhaltijaperheiden ravinnon riittävyys Suomessa vuosina 1950–1951' [Sufficiency of food in working- and middle-class families in Finland in 1950–1951]. *Sosiaalinen Aikakauskirja* 48 [Journal of Society] (Helsinki, 1954).
Saastamoinen, O., 'Multiple Views on Multiple Uses of Forests', in: L. Opas (ed.), *Finnish Forests* (Joensuu, 1997).
Sarmela, Martti, *Suomen Kansankulttuurin Kartasto* 2 [Atlas of Finnish folk culture 2] (Helsinki, 1994).
Suomen tilastollinen vuosikirja [Statistical yearbook of Finland] 1952, 1962.
Tilastokeskus [Central Statistical Office of Finland], *Kulutustutkimus* 1971, 1976 [Household Expenditure Surveys 1971, 1976] (Helsinki 1974, 1979).
Vilkuna, K., 'Jokamiehen oikeudesta' [Facts about Everyman's Rights]. *Kotiseutu* (1979).
Virtanen, A. I. & Roine, Paavo, *Ihminen ja ravinto* [People and nourishment] (Helsinki, 1956).
Virtanen, A. I. & Turpeinen, Osmo, *Tutkimus suhteellisen vähävaraisen väestön ravinnosta Suomessa 1936–1937. Tutkimuksia kansan ravitsemustilan parantamiseksi* [Relatively poor population in Finland and their diet in 1936–1937. Investigations for improvement of the nutritional state of the nation]. Komiteanmietintö [Committee report] 5 (Helsinki, 1940).
Vuolle, P. & Oittinen, A., *Jokamiehenoikeus perinteistä nykypäivään* [The Everyman's Right, Traditional and Present-Day] Reports of Physical Culture and Health 92 (Jyväskylä, 1994).

LYDIA SAPOUNAKI-DRACAKI

15 Modernizing the traditional Greek diet

The role of cookery books, 1833–1914

After the end of Ottoman rule, Greece remained a subsistence economy with little labour specialization, and a limited domestic market. The orientation of the economy remained predominantly agricultural throughout the nineteenth century and well into the twentieth. At the beginning of the 1830s, the creation of the Greek State under Prince Otto of Bavaria resulted in an immediate change in the Greek hierarchy and in the urban environment.[1] When Athens was chosen to be the capital of Greece in 1834, it underwent rapid demographic change: in thirty years the population of Athens multiplied tenfold due to its cultural and administrative functions. The new capital became a centre not only for the Bavarian entourage of the new King, but also for foreign diplomats and Greeks of the Diaspora.[2] One would see people from the provinces, the captains (leaders of rebel bands during the Greek Revolution) and notables (rich land-owners during the Ottoman rule and often, tax collectors) who wanted to be close to the royal court and the government. Here were people with high incomes, merchants, the unemployed, as well as landless peasants, mainly from the mountainous regions.[3] During the first years of the twentieth century, Athens absorbed successive tides of immigration and, as a result, 40 per cent of the urban population was living in the capital.[4] This migration brought together different practices and habits (including dietary habits). During the first forty years following the establishment of the Greek State, change was initiated by the Greeks of Diaspora who sought assimilation into western countries, were imitating their habits, and in turn transmitting them to various urban groups in Athens.

In the following sections, changes in the traditional Greek diet will be analyzed to show the effect of foreign influences; in particular, to show what kind of food items were accepted into the Greek diet during 1833–1914 via the influence of cookery books. Cookery books are a useful source whether we want to analyze this kind of foreign impact on the national diet or want to reconstruct the everyday diet. However, when trying to find evidence of dietary change based on this kind of material, it should be borne in mind that cookery books primarily reflect a wish to influence the diet. Much less can be said about cookery books actually influencing everyday cooking and real dietary change since the first data concerning urban family budgets in Greece cannot be found until 1957–1958.

Municipal tax records of 1833 to 1914 reveal that the diet of a town-dweller in Greece consisted mainly of cereals, flour, bread, vegetables, fresh legumes, fresh fruit, milk, unripened cheese, salt, and some additional minor food items.[5] Bread was the primary year-round staple. This is the reason why both central and municipal authorities showed interest in controlling the price, quality, and adequate supply of flour and bread.[6] At first, city bakeries supplied bread mainly to the mobile population (unsettled people, like sailors, without a permanent home). After 1870 it seems that urban families also bought their bread increasingly from bakeries. At the end of the century Greece and Italy were regarded as the countries with the largest consumption of bread in Europe. In the late nineteenth century, consumption of bread per capita was about 700 g per day. But only an estimated 3,000 kg of bread was produced in bakeries (as opposed to home-made). The low demand for bread production did not justify the cost of introducing mechanical methods of bread-making which explains why it continued to be produced manually.[7]

Olive oil and olives have always been a basic dietary component of the Mediterranean people. According to the Act of the Agricultural Advisers in Evia (Greece) in 1853, middle-class agricultural families consumed 5 g of olive oil per day per person, whereas twenty years later an Albanian worker used 8 g per day. On board ships, according to the diet of the Greek Royal Navy (1833), sailors consumed 7.5 g of olive oil per day and 390 g of olives weekly. However, domestic olive oil production could not satisfy domestic demand, because the methods of pressing olive oil were in general antiquated until the end of the century.[8] Wine and spirits were inexpensive, and could therefore be bought and consumed by all income groups. The traditional wine-making process called for pine resin to be mixed directly into the wine as a preservative.[9]

In the 1830s and 1840s, an abundance of food was sold in Greek towns. At Piraeus, for instance, the minutes of an 1840 municipal board meeting included a list of food items available in the city: cheese, butter, potatoes, fish, beans, chick-peas, broad beans, lentils, chestnuts, wheat, barley, olives, onions, melons, watermelons, lemons, oranges, chicken, eggs, vegetables, rice, almonds, hazelnuts, salt-cured fish, sardines, fresh and smoked mackerel, fish roe (red as well as black), pastrami (a lean spice-cured meat), sugar, and coffee. Relatively abundant was also the market of Syra, where the wealthy bourgeoisie could buy fish roe, salt-cured food, colonial products, etc., but mostly a large variety of fruit and vegetables.[10] In the same market, the British Consul recorded that quantities of beef and mutton were available, the latter being consumed mostly in winter. The Consul mentioned that the quality of meat was inferior to that available in his own country, and that goat and mutton had a strong odour, which restricted them for lower-class consumption.[11]

Advertisement of a wine factory from the turn of the twentieth century.
Source*: From the collection of advertisements of Hellenic Literary and Historical Archive (E.L.I.A.).*

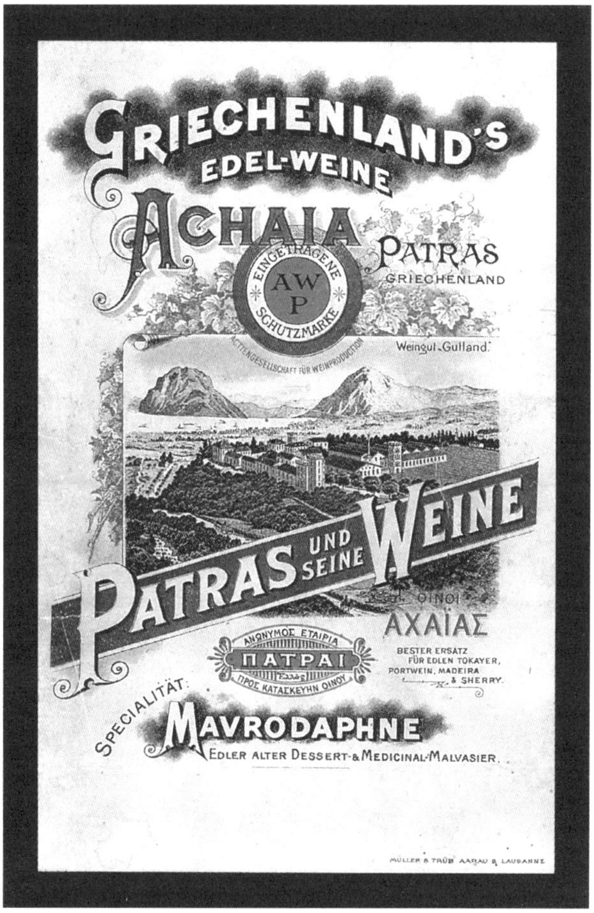

Since the earliest years of national independence, certain luxury products such as coffee, chocolate, soft *Pain d' Espagne* cake and biscuits, and beverages imported from other parts of Europe were widely consumed in Greek cities, even by people of moderate means. The Pavlidis chocolate factory produced several different types of chocolate and cocoa. Demand for these products increased so rapidly that by 1870 the factory had imported steam-powered equipment to renew and expand its plant.[12] Tea on the other hand was never particularly popular in Greece. Yet as early as 1850 one could purchase excellent quality teas at the café Anatoli, including tea imported from Russia. Greece exhibited its agricultural products at the international exhibitions of the late nineteenth century: grain, cereals, rice and peas, lentils, beans, potatoes, onions and other vegetables, as well as nuts, were all shown at Paris in 1878 along with a rich variety of fruit (oranges, raisins, figs, olives, limes, etc.). New items such as the white grapes, Raisin de Smyrne (rasalin), were also on display.[13]

First attempts to create a national Greek haute cuisine

The fundamental differences between Greek and European dietary habits were very obvious to foreigners visiting Greece during the nineteenth century. In 1830, the Bavarian, Thiersch, noted that cooking methods used in Greece for rice, vegetables and poultry, were unknown to other Europeans.[14] The first attempt to change the gastronomic customs of modern Greeks resulted from the translation of an Italian cookery book in Syra as early as 1828. It included Italian, French and other, mainly European, recipes, which introduced new ingredients and ways of cooking.[15] The first cookery book as such in Greek was published by N. Sarantis in 1863 in Constantinople and republished in 1886 in Athens.[16] The book was written for professional cooks and the Greek bourgeoisie. Besides the menus and the recipes, the book included guidelines concerning the ingredients of the dishes, methods of cooking, hygiene and cleanliness. The book also contained a chapter on the "philosophical history of cooking", which at many points reproduced the ideas of Brillat Savarin.[17] Most of the recipes were French, few of them were Greek, but none were Turkish. There are two probable explanations for the Turkish recipes being excluded: either those recipes were so familiar to the public (for instance in the patisserie catalogue of N. Pavlidis, Turkish pastry is not mentioned), or the author was trying to create a national Greek haute cuisine within the newly founded State. The latter can be seen from the archaic or heroic names that the author used to denote the dishes of his Greek haute cuisine,[18] or from the patisserie decoration, which was presented in national colours. For example, the *Greek Charlotte* showed the national colours with its waffles covered by a *Bavaroise* (Charlotte Bavarian cream) of white and blue sweet cream.

Sarantis' enthusiasm in creating 'Greek haute cuisine' led him to describe ingredients as having colours close to reality, as for example, the bassfish *à la Athenaea*, which Sarantis wanted to be cooked so that "the fish should remain white and the cucumbers did not lose their colour – as a result the white of the fish and the *mavi* (Turkish name for blue) of the cucumber would present the Greek colours".[19] In Sarantis' cookery book a number of complicated soup recipes of various origins were mentioned. Most of them were made of many ingredients in different combinations. Besides the sauces, there were also some recipes less complicated, such as oyster *à la Greca*, which was cooked on a grill with olive oil and lemon, or fried codfish. Most of the dishes were of meat or poultry. Beef, whether stewed, roasted, or as burgers was used for 35 per cent of the recipes, while another 20 per cent were poultry. In addition, 10 per cent of the dishes contained veal or lamb, and a further 10 per cent were pork. Less than 20 per cent of the recipes were based on fish. This clear majority of beef dishes can be explained by the fact that beef was considered healthy and nutritious, despite the fact that the author of the cookery book knew that the older generations preferred veal. Referring to antiquity and the dietary habits of previous times, the author also mentioned as healthy ingredients lamb and pork which, he concluded, were good mainly for winter. Pork was considered delicious, but at the same time harder to digest.[20] Among the main dishes there were only a few recipes that were *à la grecque*, i.e. everyday dishes. These are still popular today: lamb on the spit *à la grecque*, pork on the spit *à la grecque* (stuffed with raisins and pine seeds) and stuffed turkey, which has been served in Greece at Christmas until the present day. At the end of Sarantis' book there was a patisserie guide, which included mainly European recipes. In addition to these, some new non-traditional recipes appeared using ingredients of foreign origin, but they never succeeded in getting established in Greek society.

'Second generation' cookery books for a broader urban audience

From the end of the nineteenth century onwards, particularly at the beginning of the twentieth century, simpler cookery books started being published; cookery books that were straightforward, analytical, easily adaptive, and written for Greek cuisine. These books were meant for the local urban cuisine and the broader strata of the population.[21] To these belonged the *Elliniki Mageiriki* (Greek Cookery)[22] by Anestis Konstantinidis, which claimed to be a very useful guide. It included a balanced diet with Greek, Turkish, and some European dishes, using Greek cooking terminology. Except for soup, sauce, croquettes, and mashed potato recipes, the cookery book included 183 recipes, which can be categorized as follows in Table 15.1. Meat, for which there was the biggest percentage of recipes, is further sub-divided in Table 15.2.

Table 15.1 *An analysis of the recipes in Greek cookery books.*

Cookery book	Meat (%)	Game or poultry (%)	Fish (%)	Other[1)] (%)	Eggs (%)	Total
Elliniki Mageiriki 1891 [Greek Cookery]	36	16	18	24	6	100
Oikogeneiaki Mageiriki (1900) [Family Cookery]	33	27	17	17	6	100
Odigos Mageirikis (1910–1911) [Cookery Guide]	37	19	14	24	6	100

Note: [1)] Recipes with legumes, vegetables, spaghetti, rice etc. without meat.

Table 15.2 *Varieties of meat in Greek cookery books (excluding game).*

Cookery book	Beef and veal (%)	Lamb and sheep (%)	Pork (%)	Total
Elliniki Mageiriki (1891) [Greek Cookery]	58	25	17	100
Oikogeneiaki Mageiriki (1900) [Family Cookery]	49	39	12	100
Odigos Mageirikis (1910–1911) [Cookery Guide]	43	35	22	100

The common feature of the cookery books at the turn of the twentieth century was that they included a great number of vegetable dishes. This was perhaps due to salad recipes being almost absent from cookery books and weekly menus, where they appeared once or twice per week. It seems that salads did not belong to the diet of Greek urban families, even in the twentieth century.[23] *Elliniki Mageiriki* also included a broad range of Lenten dishes, which had as their basis vegetables, legumes, pasta, and even seafood or Lenten fish (such as oyster, cuttlefish, etc.), which were prepared with olive oil. If we assume that the upper social class in towns rarely fasted, it leads to the conclusion that the cookery book was meant for the broader urban classes of the period.[24] The majority of the sauce recipes in the Greek cookery (37 per cent) had olive oil as a basis (even Espagnol sauce), and 30 per cent of them were prepared with butter. Also some of them could alternatively be prepared, according to the cookery book, with olive oil as shown in Table 15.3.

Table 15.3 *Number of sauce recipes using butter, oil and fat in Greek cookery books*[1)].

Cookery Book	Butter	Oil	Fat[2)]	Milk	Cream	No Fat	Total
Elliniki Mageiriki (1891) [Greek Cookery]	5 30%	6 37%	2 14%	– –	– –	3 19%	16 100%
Oikogeneiaki Mageiriki (1900) [Family Cookery]	26 36%	13 18%	26 36%	4 4%	2 2%	4 4%	75 100%
Odigos Mageirikis (1910–1911) [Cookery Guide]	9 40%	5 22%	4 18%	3 15%	– –	1 5%	22 100%

Note: [1)] Some sauces included both olive oil and butter.
[2)] Fat (lard etc.).

Spices and aromatics mentioned in *Elliniki Mageiriki* were basically salt, pepper, nutmeg, oregano, basil, bay, and parsley. In almost every sauce onion was used, sometimes garlic, but rarely mustard. Fifty dessert recipes were included: in addition to traditional pastries one could find stewed fruit, marmalades and very sweet fruit-based desserts served by the spoon and always accompanied by a glass of water. Among the patisserie recipes there were also some sweets for religious fasting days made of olive oil, semolina, flour, and *tachina*. At least 20 per cent were of Turkish origin, and just 10 per cent were English puddings, sponge cakes, and Dutch creams.

A second cookery book in the period after 1890, by V. Alexiadis, demonstrates that by the 1890s cookery books, in order to broaden their readership, had to present recipes that were simpler and mainly shorter for Greek cooking habits. Alexiadis' book was published in 1899, and republished one year later, without the philosophic and historical texts, and without the Greek dishes of the national haute cuisine the author had taken from the first cookery book of Sarantis. The book included mainly dishes adapted to Greek taste.[25] However, it is curious that in this new cookery book of 1900, *Oikogeneiaki Mageiriki* (Family Cookery), there were no recipes concerning traditional legumes, such as bean soup, or dishes with olive oil, such as peas cooked with olive oil and onions. According to Alexiades the latter was the most common dish in Greece during that period; perhaps the recipe was, however, left out of the book because it was so well known to Greek readers.[26] In *Oikogeneiaki Mageiriki*, 33 per cent of the recipes were for meat, 27 per cent poultry and game, 17 per cent fish, 17 per cent various other dishes, and 6 per cent egg-based recipes (see Table 15.1). Of the meat recipes almost a half still concerned beef and veal, but the amount of lamb-based recipes had increased to 39 per cent (see Table 15.2). From the previous cookery book, *Elliniki Mageiriki,* the percentages changed also for poultry which increased significantly from 16 to 27 per cent compared to other main groups. This can be partly explained by inclusion of game dishes.

The influence of foreign cooking on *Oikogeneiaki Mageiriki* is obvious when we look at the combination of the sauces in Table 15.3. Most of them (36 per cent) included butter[27] or sardine butter, and the same percentage fats (lard or meat and fish broth). Compared to the 1891 cookery book olive oil was to be used less, only in 18 per cent of the recipes. Milk and cream were rarely used. Sauces included cucumbers, capers, and aromatics, such as thyme, bay, clove, nutmeg, rosemary, raisins, as well as onions and sometimes garlic. Cold appetizers were a special category in both cookery books of Alexiadis, which were known both to the diets of the twentieth century and the traditional *ouzeries*[28] of the period. Among cold appetizers were sardines, both tinned and in barrels, smoked herrings, roe-salad, green olives, salt-cured soles, *saltsiolo* (a sort of pork), *hoiromeri* (ham), oysters with lemon, and tuna fish.

Towards everyday cookery books

The best-known aid to the twentieth-century housewife, the cookery book *Odigos Mageirikis* by N. Tselementes, started as a periodical in 1910. The success of the magazine was such that not only was it republished five times as a book before 1930, but the name *Tselementes* also became the synonym for the word 'cookery book' in the twentieth century. In this magazine, cooking and gastronomy began to be studied again from an historical and philosophical perspective. The aim of the magazine was to promote healthy and digestible food (through the art of cooking), nutritious dishes (thanks to their good quality), and economical ones, without any waste of food items (through the right combinations and preparation of dishes). Also "moral good" was a basic aim of the magazine, since a wife's culinary education prevented men from going to hotels, and restaurants at the expense of family life.[29] Most importantly the magazine diffused foreign studies and culinary ideas from all over Europe. The articles of the magazine can be divided into three themes:

- scientific studies that considered, for example, definitions of taste, function of taste, sensitivity of taste, impact of smell upon taste, etc., by reference to the opinions of great men of the period and also based upon historical data;
- technical studies concerned with cooking procedures, for example, the ways in which poultry should be stewed according to some recipes; methods by which frying should be done; ways of cooking meat; and also the anatomy (with drawings) of the various cuts from different animals, etc.; and
- housekeeping studies that discussed cooking utensils, the kitchen and its arrangement, the preliminaries of a meal, and also the etiquette of meals.[30]

Much explanation in the magazine was due to the attempt to incorporate unknown dishes into the everyday diet of the readers. In the case of five new egg recipes from France, the three ways of preparation used in Greece were also mentioned. When a white sauce for asparagus was introduced, information about this luxury item – almost unknown to the Greeks – was given at the same time.[31] The recipe of *Knok's Soufflé* commented that the dish was about "*Knödels*" (German for dumplings), which were easy to prepare and common in Austria and Germany. International influences created variations of these recipes in Greece.[32] This was due to two factors: first, the scarcity of ingredients meant there was a need to use substitutes and second, that local practices required the adaptation of the recipes to conform to traditional gastronomic tastes.

A good example of the use of substitutes is a dish called game birds in crème fraîche that included cream and saffron. If cream was not available, milk soured with lemon or white vinegar could be used, and saffron could be omitted. On the other hand, to promote traditional Greek practices, the readers of the magazine were given the opportunity to contribute their own local recipes from various parts of Greece and Asia Minor (mainly from Smyrna).

These recipes included e.g. meat-pie of Cefallonia, *skordostoubi* (egg plant cooked in tomato and garlic) from Zante or *halvas* (dessert made from semolina with – optionally – almonds and sesame) from Moreas. Through his policy Tselementes preserved Greek traditions, and at the same time enhanced the newly formed urban cuisine. He urged the Greeks – even those outside Greece – not to forget the traditional Christmas sweets: *kourabiedes* (shortbread snow biscuits with almond filling), *diples* (a thin twisted hard fritter made from a batter mixture of walnuts, honey and cinnamon), *melomakarona* (honey dipped biscuits with walnut filling), milk-pies, and *galaktoboureka* (rice custard in filo puff pastry).[33]

Advertisement of a biscuit factory from the turn of the twentieth century. Source*: From the collection of advertisements of Hellenic Literary and Historical Archive (E.L.I.A).*

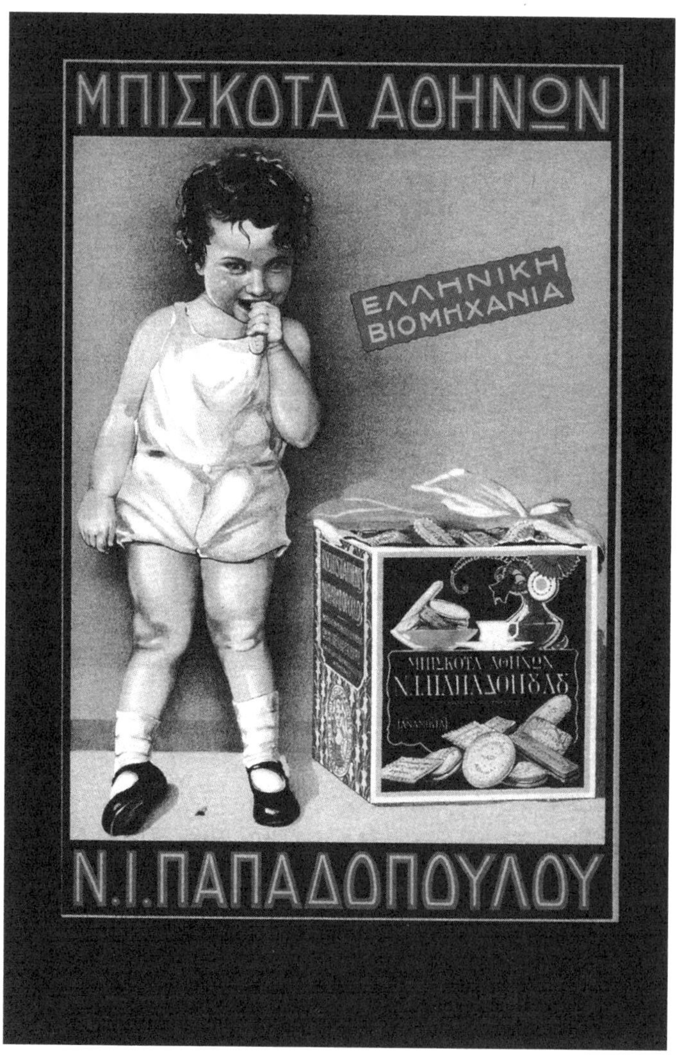

In the nineteenth century great importance was given to beef gravy and chicken broth. It was believed that these meat-based juices provided special nutritional benefits to the sick and the elderly which explains their frequent appearance in menus. Although this belief diminished, according to the physiologist who collaborated with the magazine, broths still represented a large percentage of recipes at the beginning of the twentieth century.[34] In the *Odigos Mageirikis*, 37 per cent of the recipes had to do with meat, 19 per cent with poultry and 14 per cent with fish, while legumes, pasta, and various other food items made up 24 per cent and eggs six per cent of the recipes (see Table 15.1). Again the majority of meat recipes contained veal and beef (43 per cent), while lamb and sheep were used in 35 per cent and pork for 22 per cent of the meat dishes (see Table 15.2).

Odigos Mageirikis estimated the various sauces from two different points of view: technical, in order to assure the right combinations, and gastronomic, since without sauces the dishes would be tasteless and monotonous. Mayonnaise, tomato sauce, Espagnol, and Béchamel, were the basic sauces that could be combined even with traditional Greek dishes. "Just as the painter makes infinite combinations with colours, the cook can prepare different dishes with sauces."[35] However, alternative ways to prepare everyday sauces economically were described, for instance, Espagnol was made in two ways, depending on the money available. Among the 22 sorts of sauces that were mentioned in the book, 40 per cent included butter, 22 per cent olive oil, 18 per cent broth or lard, 15 per cent milk, and five per cent fat-free food items, like sauces of English origin or sauces for poultry (see Table 15.3). The spices and aromatics used for sauces were analogous to those used in the recipes of the previous cookery books.

Finally, the magazines also included model weekly menus, some of which were repeated in subsequent publications throughout the twentieth century. In a set of ten weekly menus, for two meals per day, meat dishes were most extensively recommended. It is worth noticing the absence of legumes which, as we know, were consumed at least once per week by the middle-class urban families, even during the 1950s. They were replaced by croquettes with rice, *Knödels*, (dumplings), asparagus with cream, and other foreign dishes. Soups contributed only to six per cent of the meals, while in the diets of the previous century they had constituted a necessary complement in everyday menus.[36] Finally, 50 per cent of the meals included a dessert, mainly stewed fruit, yoghurt or *rizogalo* (milk with rice). Pastry was mentioned every Sunday and almost every Thursday. Fruit with cheese was mentioned just once.[37] The trend towards a greater number of recipes being available was apparent by the turn of the nineteenth century. This started a change in the nutritional habits of the population; in other words, a battle of gastronomy versus dietetics. Soon, articles about obesity began to be published, as well as those about vegetarian foods. Vegetarianism was suggested as a healthy diet and, by 1911, menus for vegetarians started being published.[38]

Conclusion

Most of the recipes in the cookery books and periodicals contained meat, poultry and game, while fish was covered to a lesser extent.[39] Among meat dishes the majority was beef and veal, which started appearing, especially as minced meat, in recipes at the beginning of the twentieth century. On the other hand, there were fewer recipes containing lamb (even though some change occurred), or pork, perhaps due to its dependency on the seasons. However, recipes with pork by-products like ham and salami appeared in the Greek diet either in the form of a cold appetizer, or as complementary items within the recipes.[40] Although the use of meat and fish remained at the same level in recipes in cookery books and periodicals during the whole period researched, the number of recipes containing poultry increased a little. This can be attributed to foreign influence. All cookery books gave great attention to sauces. What can be observed is that the less traditional the cookery book was, the smaller was the percentage of recipes using olive oil. At the turn of the twentieth century the nationally and traditionally important olive oil was replaced in foreign sauce recipes by butter and fats, mainly lard, which at that time became dominant in Greek cookery books. Spices and aromatics were used in the recipes of every cookery book, not just for taste but also for their nutritional values and therapeutic properties. All recipes used small amounts of aromatics, because there was a problem with the taste of the dishes. New spices of foreign origin were *karyk*, i.e., yellow pepper from India, as well as saffron which was sold at pharmacies.[41] Finally, the development of patisserie was limited. Foreign, mainly French recipes were adopted by the patisserie shops in cities after 1870, basically via confectioners who had learnt their trade in Paris. The same confectioners adapted their Turkish recipes to the European gastronomic models.[42] At that time the pastry in the cookery books was mainly European, while in the new generation of cookery books, Ottoman patisserie recipes multiplied. What should be emphasized, however, is that in everyday diet patisserie consumption was low; Greeks ate mainly stewed fruit as a dessert.[43]

The impact of foreign cuisines occurred first in the taste of the higher economic groups of the urban population, initially via translations of foreign cookery books, and later via Greek cookery books with foreign recipes. The first generation of cookery books reflected strongly the introduction of foreign dietary habits, without taking into consideration the differences existing between dietary habits in Greece and those in other European countries. This can be seen in the complexity of the recipes and ingredients used, as well as in new ways of preparing dishes. From the end of the nineteenth century onwards, a new generation of cookery books for the broader strata of the city population appeared. They adapted recipes to the Greek taste. Finally, periodicals and magazines can be defined as a third generation of cookery books, which brought new recipes more efficiently within the reach of a wider circle of urban cooks.

NOTES

1. The revolution for Greek independence began in certain regions of the Ottoman Empire in 1821. After the Turkish defeat at the battle of Navarino (1828), the treaty of Adrianople 1829, which ended the Russo-Turkish war in favour of Russia, recognized Greek autonomy. The independent Greek State was established by the London Protocol of February 1830. The first Greek Governor, Kapodistrias found himself in great internal turmoil and was assassinated in 1831. The three 'Patron Powers' then established a monarchy in Greece, and agreed upon Otto, prince of Bavaria, to be the King of the Greek State. See Svoronos (1972), 72–73.
2. 'Diaspora' is the culturally homogeneous, but socially and geographically scattered group of Greeks during the Ottoman domination, who settled over a wide area inside and outside the Ottoman empire. They moved in large numbers to Romania and along the Black Sea coast, etc. For more details, see Dertilis (1985), 55–60, 77–80.
3. See details on the demographic development of Greece in Evelpidis (1950), 80–98.
4. Tsoukalas (1975), 167–8; Loukakis (1985), 89–96.
5. In the middle of the nineteenth century the typical rural family in Evia (Greece) spent approximately 78–86 per cent of its overall budget on food. In Athens, during the decade of 1870, a typical Greek builder spent around 67 per cent of his income on food whereas a worker spent 58 per cent. See Sapounaki-Dracaki (1990), 326, 330, 374–375.
6. On market inspection, see Flogaitis (1890), 213–220; Sapounaki-Dracaki (1994), 280–283; Loukos & Samiou (1991), 62–68.
7. Sapounaki-Dracaki (1994), 279. The bakery shops in Greece from the nineteenth century until the beginning of the twentieth century made bread by hand. Only in the period of 1930–1934, were mechanical bakeries introduced in the majority of the Greek cities. See C. Zarkada-Pistioli (1994), 397–413.
8. The better nutrition on board ships was of similar standards as that of the towns as it included weekly consumption of meat which was rarely found in the diets of the rural population. Sapounaki-Dracaki (1990), 336, 425; Mathas-Demathas & Sapounaki-Dracaki (1996), 435–441. See also the Introduction by K. Spiliotakis in *I Statistiki tis georgias*, (1864),17.
9. In 1840 one could already find a great variety of wines in Athens. Starting from 1850 the Greek wineries started employing French enologists and the quality of Greek wines improved. See Sapounaki-Dracaki (1990), 179–181.
10. The vegetable market of Syra supplied a big variety of vegetables, mainly garden produce, which were exported to Constantinople. See Loukos & Samiou (1991), 86–87.
11. British Parliamentary Papers, LXVIII (1871), *Report by Consul Lloyd on the condition of the industrial classes in Syra*. According to a valid Greek economic journal of the nineteenth century, the average per capita consumption of meat, in the Greek cities was 162 g per day. *Oikonomiki Epitheorisis* 3 (25) (1875), 40-48.
12. See the Introduction by Pavlidis in *Glykismatokopoeion S. Pavlidou kai Syn en Athinais 1876* (1991), 1–34, 61.
13. Lamarre et De Queux de St-Hillaire (1878), 257–261.
14. Zografos (1923), 203.
15. See the introduction of A. Mathaiou "Oi syntages enos giatrou" in *Mageiriki anonymi metafrasi tou 1828* (1989), V–XXV.
16. See the analytical article on Greek cookery books in the nineteenth century written by Mathaiou (1998), 109–122, in which, apart from other useful information, this first Greek cookery book is analyzed thoroughly.
17. See ibid., 113–114.
18. According to the opinions of Mathaiou (1998), 114–116.
19. Sarantis (1886), 150–151.
20. See ibid., 192–203, the remarks of Sarantis relative to the various species of meat.
21. See comments about the Tselementes guide (*Odigos Mageirikis*), in the 4th edition of

the guide Tselementes (1929), 6–7.

22. See the cover of *Greek Cookery*, Konstantinidis (1891).
23. Unlike fruits, some vegetables like carrots were introduced relatively recently to modern Greek diets. According to the medical hygienist A. Katsigera (s.a.), 46–47, this was the main reason for the non-integration of these products to Greek taste.
24. Over half the year was designated as religious fasts during which all meat, dairy products and eggs were prohibited. In rural areas and sometimes in the cities, during periods of fasting, consumption was limited to bread, olives and wine. See Psyhogios (1987), 119–120; British Parliamentary Papers LXVI (1870), *Report by Consul Gould on the artisan and industrial classes in Greece*; Loukos & Samiou (1991), 86; *Oikonomiki Epitheorisis* 3 (25) (1875), 40–48.
25. See Mathaiou (1998), 118–120 for remarks on the editions of the cookery books in Greece, which aimed at the petit-bourgeois audience.
26. Alexiadis (1899), 879.
27. Tselementes (1910a), 6; Alexiadis (1899), 879.
28. *Ouzeries* are small taverns serving cold appetizers with ouzo, a drink resembling the French pastiche.
29. Tselementes (1910a), 1–2.
30. Tselementes (1910c), 88–89, Tselementes (1910f), 193–198.
31. Tselementes (1910b), 39.
32. See remarks of the author concerning the internationalization of roast beef in Greece, Tselementes (1910d), 175.
33. Tselementes (1910e), 208.
34. Menegakis (1910), 70–71.
35. Tselementes (1910a), 6.
36. Boiled meat, according to Tselementes (1910a), 12, was the Greek national dish, because lamb was expensive and therefore consumed only on special occasions. In the countryside, soups were the only daily main dish (soup twice a day with vegetables and legumes). On Sunday, villagers were eating soup, with meat, as an anonymous article in *Physis* 20 (1903), 328, mentioned. The regimen in schools also included soup for dinner. Sapounaki-Dracaki (1990), 148.
37. According to the report of a medical hygienist, Greeks did not eat much fruit, especially poor people. Katsigera (s.a.), 40–41.
38. Anon in *Physis* 20 (1903), 328–329; *Odigos mageirikis* 19–21 (1911), 409–410, 425–426, 441–442; see also the article of Dr Tavitian in *Odigos mageirikis* 19–21 (1911), 143–144.
39. Loukos & Samiou (1991), 80–81, 86, 89. For the high consumption of meat and the low consumption of fish and seafood in Syros in the later part of the previous century see also Fragkidis (1975), 96, and Fragkidis (1894), 45–46 in Loukos & Samiou (1991), 81, 89.
40. Alexiadis (1899), 7–10.
41. Platis (1892), 12–13; Alexiadis (1899), 878–882; Alexiadis (1900), 10–12.
42. See the Introduction of Pavlidis in *Glykismatokopoeion S. Pavlidou kai Syn en Athinais 1876*, (1991), 9. See also pastry making in the introduction to the first pastry-making guide, which was written by two students of Pavlidis, Alexiou & Papageorgiou (1876).
43. See note 37.

BIBLIOGRAPHY

Alexiadis, V., *Megali Oikogeneiaki Mageiriki kai Zacharoplastiki* [Big family cookery and confectionery] (Athens, 1899).

Alexiadis, V., *Oikogeneiaki mageiriki itoi leptomereis praktikai odigiai pros kataskevin fagiton epi ti basi olon ton ekdothenton megalon mageirikon xenon kai imeteron* [Family cookery namely detailed instructions how to prepare food according to all the foreign and Greek cookery editions] (Athens, 1900).

Alexiou, A. I. & Papageorgiou, P., *Akriveis odigiai tis zacharoplastikis technis ekdidomenai to proton en Elladi* [Precise instructions in the art of pastry cooking published for the first time in Greece] (Athens, 1876).

Anon., 'Trogomen ypervallontos' [We eat excessively]. *Physis* 20 (1903), 328–329.

British Parliamentary Papers (Accounts and Papers), LXVI (1870), *Report by Consul Gould on the artisan and industrial classes in Greece.*

British Parliamentary Papers (Accounts and Papers), LXVIII (1871), *Report by Consul Lloyd on the condition of the industrial classes in Syra.*

Dertilis, George, *Koinonikos Metashimatismos kai Stratiotiki Epemvasi 1880–1909* [Social Change and Military Intervention 1880–1909] (Athens, 1977).

Evelpidis, Ch., *Oikonomiki kai koinoniki istoria tis Ellados* [Economic and social history of Greece] (Athens, 1950).

Flogaitis, Theodore, *Odigos ton dimosion kai dimotikon ypallilon* [Guide for public and municipal servants] (Athens, 1890).

Fragidis, A., *I nisos Syros ypo topografikin, klimatologikin kai istorikin epopsin* [The island of Syros from the topographical, climatologiacal and historical point of view] (Hermoupolis, 1894).

Fragidis, A., *Istoria tis nisou Syrou* [The history of Syros island] (Athens, 1975).

Glykismatopoieion S. Paulidou kai Syn, en Athinais, 1876 [The Sweetshop S. Pavlidou and Co in Athens, 1876], reprint (Athens, 1991).

I Statistiki tis georgias [Agricultural statistics] (Athens, 1864).

Katsigera, Anna, *Trofi kai zoi* [Nourishment and life], 2nd edn (Athens, s.a.).

Konstantinidis, Anestis, *Elliniki Mageiriki* [Greek Cookery] (Athens, 1891).

Lamarre et De Queux de St-Hillaire, *La Grèce et l'exposition de 1878* (Paris, 1878).

Loukakis, Pavlos, 'Athina 1830–1940' [Athens 1830–1940], in: *Actes du colloque international d'Histoire, Association des études néohelléniques. La ville néohellénique héritages ottomans et Etat grec* (Athens, 1985), 85–96.

Loukos, Christos & Samiou, Dimitra, *Oikonomikes symperifores, psychologia kai viotiko epipedo enos syrianou tokisti: Stefanos D.Rigas* [Economic behaviour, psychology and living standards of a usurer from Syra: Stefanos D. Rigas] (Athens, 1991).

Mageiriki anonymi metafrasi tou 1828 [Anonymous cookery book translated in 1828] (Athens, 1989).

Mathaiou, Anna, 'Vrigia Savarin kai to psari tou Archinavarchou Kanari: Ta ellinika vivlia mageirikis tou 19ou aiona' [Brillat Savarin and the fish of chief admiral Kanaris: the Greek cookery books in nineteenth century Greece]. *Historica* 15 (28–29) (1998), 109–122.

Mathas-Damathas, Zacharias & Sapounaki-Dracaki, Lydia, 'To ladi stin Ellada tou 19ou aiona: Katanaloseis kai times' [Olive oil in nineteenth century Greece], in: *Proceedings of the 4th workshop, the ETVA Cultural Foundation, "Olive and Olive Oil", Kalamata, 7–9 May, 1993* (Athens, 1996), 432–450.

Menegakis Ioannis, 'Epistimonikai meletai Zomos kreatos' [Scientific researches, meat broth]. *Cookery Guide* 3 (1910), 70–71.

Odigos mageirikis 19–21 (1911), 'I fytofagia kai kreofagia' [Vegetarianism and meat-eating], 143–144, 409–410, 425–426.

Oikonomiki Epitheorisis 3 (25) (1875), 'H nisteia apo oikonomikin epopsin' [Fasting from the financial point of view], 40–48.

Platis, Stamatios, *Megali mageiriki kai zacharoplastiki diirimeni eis tessera tmimata* [Big cookery and confectionery divided into four parts] (Piraeus, 1892).

Psychogios, Dimitris, *Proikes, foroi, stafida kai psomi, oikonomia kai oikogeneia stin agrotiki Ellada tou 19oy aiona* [Dowries, taxes, currants and bread, economy and family in nineteenth century agricultural Greece] (Athens, 1987).

Sapounaki-Dracaki, Lydia, *La Grèce urbaine et rurale (1835–1875): consommations et revenus* (Paris 1 Ph.D., 1990).

Sapounaki-Dracaki, Lydia, 'Timi kai poiotites tou psomiou: Kratiki paremvasi kai dimotiki politiki sta ellinika astika kentra (1839–1900)' [Prices and qualities of bread, state intervention and municipal policy in the Greek urban centres (1835–1900)], in: *Proceedings of the 4th workshop, the ETVA cultural Foundation "Our bread from wheat to loaf", Pilio, 10–12 April* (Athens, 1994), 278–299.

Sarantis, Nikos, *Sygramma tis Mageirikis,* [Treatise of cookery] (Athens, 1886).

Svoronos, Nikos, *Episkopisi tis Neoellinikis Istorias,* [History of modern Greece] (Athens, 1972).

Tselementes, N. K., *Odigos mageirikis* [Cookery guide] 1 (1910a).

Tselementes, N. K., *Odigos mageirikis* [Cookery guide] 2 (1910b).

Tselementes, N. K., *Odigos mageirikis* [Cookery guide] 4 (1910c).

Tselementes, N. K., *Odigos mageirikis* [Cookery guide] 8 (1910d).

Tselementes, N. K., *Odigos mageirikis* [Cookery guide] 9 (1910e).

Tselementes, N. K., *Odigos mageirikis* [Cookery guide], 4th edn (Athens, 1929).

Tselementes, N. K.,'Oikonomikai meletai' [Economic studies]. *Odigos Mageirikis* 9 (1910f).

Tsoukalas, Konstantinos, *Exartisi kai anaparagogi o koinonikos rolos ton ekpaideytikon michanismon stin Ellada, 1830–1922* [Dependence and reproduction: the social role of the educational services in Greece, 1830–1922] (Athens, 1975).

Zarkada-Pistioli C., 'Ta artopoieia sti Florina thn periodo toy Mesopolemou' [The bakeries of Florina in the interval between the two world wars], in: *Proceedings of the 4th workshop, the ETVA cultural Foundation "Our bread from wheat to loaf", Pilio, 10–12 April* (Athens, 1994), 397–413.

Zographos, Dimitrios, *Istoria tis ellinikis georgias (1821–1833)* [History of Greek agriculture (1821–1833)], vol. 2 (Athens, 1923).

VIRGINIE AMILIEN

16 A taste of authenticity

Nature and tradition in modern Norwegian restaurants of the 1990s

C. Levi-Strauss frequently described food as a perfect link between the concepts of nature and culture. As a matter of fact nature is playing an increasing role in the Norwegian catering landscape, and indeed this coincidence of tradition and nature concepts is not totally unexpected. Nature has always been a fundamental concept in Norway, especially since the national-romanticism movement in the 1850s, when nature in the wild used to be considered as the perfect reflection of the original Norwegian soul. Moreover the traditional culture of food is two dimensional, with the old-fashioned peasant cookery, which apparently makes Norway an un-elaborated cuisine[1] in a European context on the one hand, and restaurants' international cuisine that appeared more than a century ago on the other. In a structuralist way of thinking we should distinguish between popular traditional cuisine, often considered as natural, and professional traditional cuisine, which is built on an international cultural frame. Although restaurants' food culture has always been distinguished from the popular one, the Norwegian urban restaurant landscape is in total evolution, indeed revolution, as shown by the Bocuse d'Or 1999 (won by the Norwegian team) or the nature-food philosophy of Arne Brimi's nature cuisine.[2] From that perspective, this study of the authenticity of meals served in urban restaurants has a strategic value.

Before presenting the empirical material, which mostly consists of restaurant menus and leaflets from five of the largest towns in Norway, a short explanation of the Norwegian national identity and food history is necessary to set up the contextual values. The historical survey of Norwegian food culture and Norwegian restaurants, which follows should enable the evolution and the actual standpoint of the subject to be better understood.

Restaurants' presentations and menus are linguistically rich in details and explanations. They often quote "fresh natural berries", "traditional recipes" and, what they call, "real Norwegian cuisine". By a textual and contextual study of all these menus, it is possible to follow two parallel questions and axes: What does tradition mean in this perspective and how is nature used to justify the authenticity of the meals we eat in these restaurants?

The emergence of modern Norway

Norway is quite a young country, in terms of independence, although some people were already living in the forests and along the fjords about 8000 BC. Datable Norwegian history began around 900 AD, during the time of the Vikings, when Harold the Fairhair (Harald Hårfagre) made a kingdom from all the small regions of Norway; a kingdom that was Christianized about 100 years later. In 1380, Norway became a province of Denmark and the two countries made a single economic entity. Norway was then a country with few people (half the population died during the Black Death in 1349) and almost no urban life. The Danish kingdom monopolized the most important functions, and people working within Church, law or education were Danish. In Norway, urban language was influenced by the Danes, as were education and food ways. Thus, when the separation finally came in 1814, Norwegians had to find their own identity. They drew upon national romanticism, and when Sweden gave Norway its complete independence in 1905, were still engaged in this identity quest.

One thousand years as a political kingdom, four centuries under Danish hegemony and one hundred years of political independence, complicates the evolution of Norwegian foodways. Although Maurizio asserts that "in a food history perspective, there is almost no change within 1,000 years"[3] we have to consider modern times and recognise that the so-called post-modern world has experienced a much faster evolution.[4] A brief survey of Norwegian food history will explain first the long stagnation and subsequently the recent transformation of food habits.

"Our people ate what they could" and "the old cuisine depended on the harvest of the year" is the way Hilmar Stigum introduced his book about Norwegian food fifty years ago.[5] This could have been said as a general description of any pre-industrial country, but it remains an anthropological assessment that people believe in. The basis of the Norwegian diet was mentioned in the old Scandinavian literature and folktales: porridge and the famous flat bread existed at the time of the Vikings.[6] P. Ch. Asbjørnsen, a Norwegian folklorist, who first tried to compile a popular cookery book, wrote in 1864 that two-thirds of Norway ate porridge, both in daily life and on festive occasions. "The fatter, the better" was the expression people used about their food, whether porridge or soup or meat. Fresh food was only rarely used, and both meat and fish were salted, smoked, dried or fermented, for preservation.[7] Milk and milk products were basic foods, while berries and other natural products, like game, were occasional supplements, but never a usual part of the daily food. Food culture depended mostly on locality, and was totally different in the mountains or on the coast. Nevertheless, food culture is also a matter of social habits and everyday life, and there were huge differences between peasants and merchants. Country people used old recipes, while tradesmen ate "in the Danish way", as P. Ch. Asbjørnsen put it.[8] The Norwegian kitchen appears divided geographically, until the strategic and political unification around 1850, when the old peasant cuisine logically (with regards to national identity) became the popular one. At the same time the first city restaurants arose, influenced by the French model. This professional culture of food was built on international cuisine, and had completely different foundations and values from the popular one. On the one hand was the popular way of thinking where quantity, comfort and nature were the main values and on the other hand quality, gastronomy and culture were most important.

The social aspect of food culture was also linked both by the daily and annual rhythm, from which arose the seasonal feasts. Three meals a day seem to have been common for several centuries, though with variations between pre-industrial, industrial or post-modern societies, for example.[9] Seasonal feasts were the *rites de passage* of food culture and played a basic role in framing tradition. They could have a social meaning, as marriage, or a religious and ritual one, as Christmas. Every situation had a peculiar local food tradition, which often survived until today. It is the reason why this type of meal is often seen as the traditional food culture in Norway. They are witnesses from the past, and as such, are often used to establish a common identity. Food habits were passed orally from generation to generation. Already in 1864 Asbjørnsen knew that his cookery book would be "directed towards city houses (because they do not use cookery books in the country-side)".[10] In cities, restaurants were also used for the seasonal or ritual feasts, and eating out usually evinced a special event. Nowadays identifying with rural society is a practical problem. Post-industrial change has transformed the social landscape. Peasants now represent only a small part of the whole population: by 1998, around 25 per cent of Norwegian citizens were living in the Oslo area.

In a report about Norwegian food culture written 15 years ago,[11] Wold argued that it was defined by two main axioms: first, almost everybody ate at the same time, three times a day, and second, 76 per cent of the dishes were eaten together with boiled potatoes. A 1999 report[12] found Norwegian food habits still mainly traditional, though undergoing change. Most people still eat three times a day, and boiled potatoes are still the favourite food, as in 1985, but the younger generation and the more highly-educated sector of the population living in cities are changing progressively. Either they adopt different eating patterns or new habits like eating fresh vegetables and fruits. Change mostly depends on time conception and globalization on the one hand, and ideas about diet and health considerations on the other.

A recent comparative survey about food habits in four Nordic countries[13] shows that there is a national difference, at a practical level, both in the way food consumption is structured (four meals a day: breakfast, *matpakke* a wrapped sandwich at lunch, dinner and late evening supper) and in the type of ingredients (boiled potatoes, pasta, a few vegetables, etc.). On the other hand there does not seem to be any specific ideas of a national diet; instead people commonly say, "we eat the same as the others". Norwegian people usually refer to regional or local food when they need to specify any traditional course or meal.[14] This lack of a national concept or definition of food culture, could be explained by the fact that there was no active political discourse about food culture until the last five or ten years, apart from the *matpakke* movement[15] and a number of political decisions upon health.[16] From this perspective, there has been a major change in the discourse about food culture since the introduction of gourmet food and Norwegian food is nowadays supported by daily articles in newspapers or popular television programmes.

A parallel development is apparent among Norwegian restaurants. They have changed greatly since the 1850s, but especially in the last 15 years. Geographical, climatic and religious conditions were never conducive to Norway becoming a country of restaurants. Even nowadays, at the end of the twentieth century, most Norwegians are not used to dining out, and prefer to meet at each other's home or just to take a drink when out. Dining out is not yet a part of popular culture, and recent reports show that only one in 50 meals is taken out.[17] But consumer attitudes towards restaurants and food are slowly changing, mostly in cities, and among young or highly-educated people. This is underlined by the growing number of restaurants, which has almost doubled in less than 10 years.[18] There has been a marked change in the urban landscape, where the number of restaurants in 1997 was sometimes more than twice that in 1988. As typical of the effects of globalization, the number of fast-food outlets has increased steadily in recent years, but so have ethnic restaurants of all kinds (where the Chinese and Italian seemed to have a sort of monopoly before). There are even culinary experience restaurants. The strength of the Norwegian economy (average income is high, and people are able to travel abroad, and even to dine out in spite of very high prices) and the open European food market are the official reasons for the transformation. Moreover, it should be noted that the Department of Agriculture is actually supporting an active programme with regard to food culture, while some Norwegian cooks are becoming more and more influential, thus making the Norwegian haute cuisine an almost daily matter in the mass media. These cooks are an attraction in themselves. They are considered as the ambassadors of Norwegian traditional food.[19]

Modern Norwegian restaurants and traditional food

In our country of frozen pizzas[20] and sausages in bread, what counts as traditional food in the 1990s? How do city restaurants employ the concept? To what extent do urban restaurants make use of tradition and nature to justify authenticity? And finally, why have they chosen traditional food to signify the modernization of Norwegian eating?

Reading menus is a wonderful linguistic experience. Words, sounds and pictures interact in the world of restaurant menus for the pleasure of the consumer. The idea of this study was originally to analyze menus and the public relations information of Norwegian-food restaurants, to get an idea about the kind of food tourists, either Norwegian or foreigners, could read about and eat in the largest cities of Norway. For this reason the so-called ethnic restaurants, as well as those restaurants not accessible to tourists (either because of location or intended clientele), were not directly taken into consideration. This study sampled about a quarter of the working restaurants in five of the largest towns in Norway. From a purely numerical point of view, the empirical material consists of about 200 menus and 50 information leaflets from different restaurants.[21]

The main difference between the five towns is a contextual one. Oslo has changed a lot during the last 15 years, and has become a modern European city, with all the essential signs of globalization. The first coffee house opened at the beginning of the 1990s, and it is nowadays almost impossible to walk about without noticing a new café, where young people drink espresso and eat croissants. Almost 1,000 eating places were registered with the communal administration in 1998, which was twice of that in 1988. Of these 1,000 eating places, about 60 cafés and restaurants corresponded to our criterion – of having some Norwegian food – from simple family restaurants to the most exclusive ones. As the capital of Norway, Oslo is visited by the majority of Norwegian and foreign tourists. Restaurants, mostly concentrated in attractive areas, offer a wide range of culinary experiences. Almost all types of ethnic restaurants are represented, and even the only pure Norwegian restaurant. This "wide selection of seasonal and traditional Norwegian food" from Norwegian produce prepared in the old peasant way, makes an exception – the first but certainly not the last – in the restaurant landscape. Apart from this case, multiplicity and variety in the choice of foods, are the most specific common features of Oslo's restaurants.

In Bergen, almost all restaurants have the typical creamy fish soup, and often mussels and scampi. In spite of a rainy climate, Bergen is a perfect tourist attraction with its old houses and wonderful fjords. Its cultural programme led to Bergen being the European City of Culture in the year 2000. From a food point of view, it is the fishing centre of the country, and fresh fish or fish cakes are considered as a culinary experience.

Trondheim is the most historical city of Norway, with the Nidaros cathedral as its focus. One thousand years ago, King Olav took his army to Nidaros, built houses on the hill by the river Nidelva and laid the foundations for a town. This historical event has a great importance in a country that is still searching for its identity. In the official presentation of Trondheim, we find a description of Norwegian food: "The old, traditional dishes such as *lutefisk*, cured meats, herring, salted fish and game are all popular today. Fish has seen a rebirth on our tables. And as in the old days, beer is once again a popular drink with our meals, served alongside a small glass of Aquavit, which is made in Trøndelag!" These sentences are witness to a renaissance of food traditions, underlined by words such as rebirth or again for example. Menus from Trondheim are not necessarily so tradition-oriented as the author of the quotation insinuated, but several restaurants are working with local products, especially fish, and regional or national recipes.

Restaurants in Tromsø focus on fish. Many kinds of "fresh" fish are offered, even in a non-specialist fish restaurant. The Northern part of Norway has built a tourist image based on the wildness of nature and sea. Several restaurants of the city are members of the Arctic Cuisine programme, which promotes "culinary arts with fresh ingredients from the Green Arctic". With wild mountains and coast as a background, they present the "best and freshest foodstuffs in the world, from both sea and country". They offer local specialities, and almost all types of fish. Steaks of whale, seal or shark are available, and the ugliest fishes from the sea, cat-fish or wolf-fish are becoming "a true delicacy". Nature is becoming a cultural asset in Tromsø. The Northern part of Norway has actually kept a great part of the old food traditions, essentially based on fish consumption. Nevertheless, in a study about the Båtsfjord area in Finnmark, the Norwegian anthropologist Marianne Lien noted how fast-food culture had changed during the last 20 years. In spite of a living myth about northern traditional daily foodways, she noticed an evolution based on social, cultural and material factors.[22] But some years ago, the Norland Academy of Arts and Sciences and Trade and Business in Northern Norway began to support the Arctic Cuisine, "an encounter with North Norwegian Cuisine daily and for festivities". It includes grandmother's kitchen as well as modern cuisine, with information directed towards both local people and visitors.

The Stavanger strategy is quite different. Region and commune give political and economic support to food culture, and focuses on local products. Rogaland, the region around Stavanger, has been *the* culinary area of Norway. There is a food festival in summer, and a number of great Norwegian chefs have graduated from the *Norsk Hotelhøyskole* (Hotel and Restaurant Technical High School) in Stavanger. Originally a fishing town, Stavanger became an oil town in the 1960s with much foreign influence. This might explain the special restaurant landscape in Stavanger, where it seems difficult to find traditional Norwegian food. Restaurants offer menus more adapted to the Stavanger population, than to tourists travelling through.

The restaurants studied can be divided into three main groups:[23] Group 1: *Exclusive formal restaurants*, with excellent food, haute cuisine and high quality service. The kitchen is usually described either by "French inspiration and Norwegian touch", or by "haute cuisine française", with a menu in perfect French. Anyway, they frequently offer moose or fish from Norway. Menus often vary each day, in accordance with market availability. Price is often around 1,000 Norwegian crowns (about US$170) for a four course meal, and people have to reserve several weeks in advance. Tourists rarely eat at these expensive places, but there is a potential for them to do so. Group 1 constitutes about 10 per cent of the whole sample.

Group 2: *Quality informal restaurants*, with good food, even quite original and modern. The kitchen is often described as international, or French or even Norwegian. These restaurants almost constantly refer to "fresh food", "fresh local product", "pure Norwegian vegetables and fruits", "best Norwegian foodstuffs", or "best quality foodstuffs", to quote the most common words. Menus follow the seasons, which gives a new opportunity to underline Norwegian nature and freshness of foodstuffs. One main dish costs between 150 and 300 crowns ($25 to $50) depending on the type of food, meat being a little more expensive than fish. Group 2 makes about 55 per cent of the whole sample.

Group 3: *Convenience local restaurants*, often called cafés in Norway. Most of them (about 75–80 per cent) just have hamburgers, pizzas and sandwiches, and possibly some meatballs.[24] But some of them try to offer a kind of "daily and familiar Norwegian food", with huge portions and good grandmother-type of cooking. They usually refer to a "traditional way of cooking", with "traditional dishes", but never specify Norwegian produce or fresh foodstuffs. These places are quite small (a few tables) and often are outside the tourist circuit. The cost of a main dish is about 65 to 90 crowns ($10 to $15), which is extremely reasonable in terms of typical Norwegian prices. Group 3 covers about 35 per cent of the whole sample.

Menus are very similar from one town to the other, both in content and form, if we compare within the same category. Group 1 restaurants are representative of haute cuisine, and often use original sauces and ingredients such as goose liver or fillet of duck. Salmon and reindeer are almost never offered, though fish or game is. Desserts are often home-made sorbets or patisseries. Group 2 usually has a good choice of meat and fish, always including at least a piece of reindeer or another typical Norwegian meat, and often some salmon variation, either smoked or fresh. Fresh vegetables, rice and all kinds of potatoes accompany the meat and fish. Desserts are often composed of ice cream, pudding, fruits or cakes. Group 3 always has meatballs, pork and bacon. Meat can be salted, smoked or fresh. Everything is served with boiled potatoes, and possibly some root vegetables. Fish is rare, but one might find boiled cod or fried plaice. There is either no dessert, or *gelée* and pudding, sometimes ice cream. Quite cheap children's menus might be offered.

Common points and divergences between menus within these five cities stand out the following way: fish, especially sea fish, is obviously the common denominator of every region. Types vary a lot, but salmon is still the king of the menus, because of recent fondness for this type of fish. There has been a re-evaluation of the most common types of fish like cod, for example, due to Government policies, especially in the exclusive restaurants. [25] Game, particularly moose and reindeer, and wild berries are often presented as national specialities, and can be found everywhere, at any culinary level. Whale meat has been very fashionable in the last five years, as a local and traditional product.[26] The point seems to be that Norwegian nature is able to provide fresh and original foodstuffs, which a greater number of restaurants decided to integrate in their menus.[27] Differences are mostly contextual, and menus are actually quite similar. This is emphasized by the official Norwegian publicity on the internet:

> Magnificent landscapes, climatic contrasts and man, alone, pitted against the crushing forces of nature. All of these are part of the fascination that Norway holds for visitors from other countries. But nature and climate are more than merely the basis for memorable experiences, they are the very source of the ingredients from which Norwegian culinary traditions have evolved. Norway's long and varied coastline offers ample waters for both 'wild' and farm-raised fish. The slow ripening process of everything that grows during the light Nordic summer imparts an extraordinary aroma to

> berries, fruits and vegetables, and the animals that graze on the succulent green grass provide meat with a distinctive full flavour.
>
> Norway is a modern, urbanized nation with a booming petroleum and technology industry. However, with only four million inhabitants spread throughout an area the size of Italy, it goes without saying that unspoiled nature dominates most of the country.[28]

The essence of Norwegian cuisine

The most specific common feature of all restaurant information leaflets, is undoubtedly their systematic reference to traditional kitchens, to fresh foodstuffs and pure Norwegian products. In both text and way of thinking their publicity material echoes the Government's quotation above, which could be understood as a cultural reconstruction. Only a minority of restaurants has information leaflets; it is principally a characteristic of Group 2. Groups 1 and 3 do not really need it: exclusive restaurants live on their name, and local restaurants usually live on their food. In about 30 Group 2 restaurants in Oslo, five main references to Norwegian foodstuffs, seven to fresh products, three to food in season and three to tradition occur. These concepts may be mixed and associated with each other. In one case, on a page of about 200 words, there were six references to Norwegian (associated with best or fresh) and four uses of tradition. Other restaurants did not make any of these associations. Their information usually stressed the pleasant location, an historical value or the visit of famous people. That means that almost two thirds of Group 2 restaurants played on nature and tradition to attract consumers.

Atmosphere and cultural values are much more important in Bergen. Only six of the thirty restaurants with information pages referred to best foodstuffs or daily foodstuffs. Of course fresh fish was important in the fish restaurants, but tradition was only named in one case. French or international cuisines are often used as an attraction, but this does not prevent reindeer or moose from being on the menu. In Trondheim, fish restaurants simply refer to their food being from the day's catch. Restaurants having Norwegian dishes present them as from a traditional recipe, without any particular insistence on local products. On the other hand, restaurants in Tromsø not only focus on Norwegian foodstuffs but also on Northern Norwegian foodstuffs which easily becomes "the best in the world". Of 17 restaurants, four had a main paragraph on Arctic food, five on Northern Norwegian culture, four on fresh product from nature and one on typically Norwegian food. Interviews showed that fish is the typical food from Northern Norway, and that both Norwegian and foreign tourists expected to eat good fish in Tromsø. Game and whale meat is also quite popular (even the Slovak restaurant had it!). Stavanger restaurants are very varied, ethnically, and so are their information leaflets. They promote an exotic food experience, either French, Italian, American or Norwegian. Every place stipulates fresh and local foodstuffs, cooked in a special way.

Nature and tradition have been collective features shared by the great majority of restaurants' information leaflets. This is logical if we consider that they are the central concepts of the cultural image Norway was built around. But the disproportion between the words and the dishes is amazing: publicity material refers to survival and fresh food, while menus and kitchen are actually far from these folkloric notions. For instance: one of the restaurants in Group 2 has an information leaflet in honour of their "Norwegian summer menu, and the excellent Norwegian kitchen". It informs diners that the restaurant has picked only the best from their "treasure chamber" so that guests can get a real "taste of Norway". In addition, it has chosen "a few ingredients from the International kitchen [...] to improve the traditional and original". Nature and tradition, though almost unmentioned, are constantly underlying themes. Excellent, best, real are the adjectives chosen to support Norwegian kitchen, treasure chamber and taste of Norway. Those three concepts are indeterminable, but make a kind of sense in the collective imaginary perception of Norway.

The paradox comes when you sit at the table, and read the first side of the menu, composed of an Andalusian gazpacho and a rack of lamb provençale, alongside a poached halibut *Bygdøy* (Norwegian fish at its best!). Main courses are served with Norwegian root vegetables, red-wine sauce and boiled potatoes with a sour-cream mousse. The few ingredients from the International kitchen are making half, if not more of the menu. The à la carte menu follows the same pattern, mixing Oslo fillet of mackerel and cured ham with melon. All the so-called traditional Norwegian dishes are authenticated with an historical or cultural name, like Bygdøy (a well-known area in Oslo), Oslo, Norwegian, or Peer Gynt. The most national of all, the well known *lutefisk*, "dried cod fish, an old Norwegian delicacy that dates back to the fifteenth century", is even called by the name of the restaurant itself. They do not, however, specify that this delicacy is traditionally eaten in winter and not in summer time. In another example, after presenting the traditional Norwegian cuisine, the restaurant offers a "tuna carpaccio, marinated in olive oil, served with parmesan, olives, basil, pistachio nuts and a mild horseradish sauce", followed by a "salmon fillet turned in honey and coffee before grilling, served with asparagus beans and French beans sautés in butter" and a "rack of lamb with olives and a rich mustard hollandaise sauce". The dessert that closes this "traditional menu, made with fresh raw materials from nature's larder" is "a cherry pie and a pistachio cream sauce". These are two of the most extreme examples, but still represent the typical dualism of restaurants in Group 2. Group 1 uses international or French cuisine as their main reference, and Group 3 does not mention fresh foodstuffs or nature at all: there is no paradox. Nevertheless, in the context of Norwegian food history, the association of traditional and fresh food is itself just such a paradox.

This linguistic compilation does not make sense in culinary terms (where traditional food was anything but fresh), but implies a cultural and historical context. And this very context is more topical than ever. Tradition and nature are the intersection of Norwegian national identity, at a time when modern European cuisine is in the process of globalization, and new dietary and ecological food problems have beset western society in the last decade of the twentieth century. Nature and tradition are not only the two concepts that best awake the 'typical' romantic national feeling, but also those which appeal to special cultural features (seldom expressed in a global consumer society) and to an almost unique trust in the nation. Eivind Stø recently wrote that "the country of origin is important for Norwegian consumers, especially for foodstuffs [...] product quality is the main reason for consumer's choice of Norwegian foodstuffs [...] but a big minority buy Norwegian products in order to support farmers and national industry".[29] He also notices the fact that this patriotic attitude has grown during the 1990s.[30] The 'trust value' can explain a part of the quotations like "fresh and natural Norwegian foodstuffs", where fresh or natural have a very useful duality. But by way of conclusion let us confirm that nature and tradition are definitively a part of the renaissance of Norwegian food culture, but their role until now has been more ideological than practical.

Tradition is therefore used to mean survival, to provide a historical reference that gives cultural meaning. In restaurant perspectives, every type of food, which is a part of the collective Norwegian national identity, might be called traditional, or typical Norwegian.[31] That includes game, wild berries, fish (mostly from the sea, but recently from river and lakes), boiled potatoes, sour cream and flat bread, to give the most usual examples. Some of them have become national food, as *raspeboller*, *lutefisk* and *pinnekjøtt*[32], while others remained local specialities. Nevertheless, the romantic vision of the 1850s, based on wild nature and animals, is now the typical one. In spite of this, it could be asked why urban restaurants systematically need to refer to tradition and nature, while their cuisine alone could justify its delicacy, by taste, aroma, consistency or colourful presentation? A cultural Norwegian feature, obviously based on geographical and religious factors, could be an answer: food is a substance, not a delicacy.

Yet there are some political explanations. Norwegian food culture today is to be found in the political vision of national identity. National identity is a cultural construction that needs concrete references to survive. Nature, until now, has been the major argument used to bring together the Norwegian people. In the 1850s, the Government used the rural model to distinguish Norway from a more urban Denmark. They referred to Norwegian folktales, rural style of life and furniture, peasant clothes, with the creation of a national costume, the *Bunad*, in the 1920s, and so on. By associating national food culture with the old peasant tradition, the government has established a semantic thread that gives an authenticity to an imaginary common culture. The orientation of cultural evolution towards nature and the old peasant habits has been done in a skilful, systematic and resolute way.

Norwegian urban culture rests on a structural duality with, on the one hand, the most modern society and all its characteristics, and on the other, old peasant traditions. Apart from a handful of intellectuals, claiming their urban identity, most Norwegian city-dwellers refer constantly to the bygone past of their 'real' origins, either at the coast or in the mountains. In his *History of Skiing in Norway* Tor Bomann-Larsen showed humorously, how even the urban elites are walking around (and even eating at table, in spite of conventional manners) with a Norwegian knife hanging at the belt, as peasants did before.[33] This folkloristic invention of an old peasant attitude as the 'real' Norwegian image is well-known.[34] But it is important to notice the impact of this mentality on popular cuisine.

The basic dichotomy between countryside and city lies in this constant pursuit of a national identity, which is systematically bounded to the past and olden times. It is the myth of the golden age, which prevents the recognition of a modern and peculiar collective identity. Foodways provide one of the clearest examples of this uncomfortable duality, but can also become an easy way out of this cultural ambivalence. In this perspective, modern urban Norwegian restaurants refer to the golden age, but are already cooking in a modern manner concerned with freshness and taste.

Dietary and gastronomic trends differ overtly from one social class to another and cuisine has been, and still is, used to mark social differences. While the European social elite usually directed and fixed the new trends in matter of culinary evolution in older parts of Europe,[35] the peasant was chosen as the socio-political model for the whole society in Norway.[36] And it is obviously the reason why a lot of urban restaurants exploited the rural spirit of Norway to seduce the client, whether Norwegian or foreign tourist, who are both accepting the same myth.

The development of haute cuisine was actually a godsend for the Norwegian kitchen. Not only because some Norwegian cooks working in France 20 years ago came back to open their establishments in Norway, but mostly because of the nouvelle cuisine philosophy itself. Nouvelle cuisine is defined by four main principles:[37] always using freshly marketed food, using modern technical support, thinking about health, and showing creativity. These new rules gave foreign cooks an opportunity. Until that time, around 1970, French cuisine was based on traditional regional cookery and local ingredients. With the nouvelle cuisine, foodstuffs from all over the world can be integrated if fresh; something that modern transport systems have made possible. Fish is fashionable, especially salmon and other sea fishes. Freshness and invention are the two key words that opened the door to the New Norwegian Kitchen and the nature cuisine. A statistical study shows that concepts of freshness and best foodstuffs are five times more important than survival or tradition in restaurant information leaflets. Yet if the four principles of nouvelle cuisine are combined with the typical Norwegian features, it adds up to one of these restaurant leaflets! With a turnover of about eight milliard crowns in 1996 there are also economic reasons for Norwegian restaurants to use these descriptions of their food. Information leaflets aim to seduce the client, by creating an image of authenticity. Norway has institutionalized nature, giving it a structural frame with cultural values and a social role. Norwegian nature has been put at the service of the consumer, so that it now has its own economic value, as well as tradition. Authenticity is not a common phenomenon in tourism nowadays, but emphasizing authenticity could be an answer to the so-called MacDonaldization of the world.[38]

NOTES

1. The expression is borrowed from Mennell (1989), 280.
2. Arne Brimi, a famous Norwegian chef who is teaching nature cuisine. See for example *A taste of Norway, a cookbook based on Nature's own ingredients* (1987).
3. Maurizio (1932).
4. See for example Fischler (1933).
5. Ambjørnrud et al. (1965).
6. An Englishman described the diet as follows: "the staple food of the country folk in Norway [...] it is simply meal and water baked on a large, circular ironplate and is about the thickness of a cardboard [...] the taste for it is easily acquired in the absence of other food, and with butter it becomes quite delicious, to a very hungry man." Cit. in Anon. (1998), 25.
7. This last method of preserving food could be quite difficult for foreigners to accept. To quote another part of the book *Three in Norway, by two of them* written in 1882: "We noticed an unpleasant odour, and found that it proceeded from this barrel, the fish being in an advanced stage of decomposition [...] They were intending to have a feast as soon as it should have become rotten enough [...] Truly a Norwegian has the nastiest notions of food". Ibid., 110–111. The *raki* fish is actually a perfectly delicious recipe, in spite of a strange preparation and a potential terrible smell.
8. Asjørnsen, cit. in: Bonifacius (1864).
9. The phenomenon has, for example, been studied by Jacques Le Goff in his excellent work on time perception in Middle Ages.
10. Asbjørnsen, cit. in: Bonifacius (1864), 47.
11. Wold (1985).
12. Fagerli (1999).
13. Kjærnes (2001).
14. Amilien (1999).
15. Norwegian *matpakke* has been traditional food in the country. *Matpakke* could be defined as some slices of bread (from two to four as an average), embellished with, for example, cheese or ham or salami. The pile of bread is usually wrapped in paper, which allows it to be taken to school or work. The sandwich was initiated in 1932 as a healthy alternative to the old lunch system. Runar Døving has written an excellent article on *matpakke*, showing how this quite recent habit had been the traditional Norwegian lunch. Døving (1999).
16. Døving (1999); I. J. Lyngø has written about vitamin politics in the 1930s and its influence on Norwegian foodways. Lyngø (1999).
17. Bugge & Døving (2000).
18. See *Statistics on Travel*, Oslo-Kongsvinger (1988 & 1997).
19. And they actually are more than representative, as they made their own definition of traditional food: "food, which is at least 1,000 years old". (Det Norske Kjøkken).
20. One of the highest frozen pizza consumption per inhabitant in the world.
21. Norway is a large country with an average density of only 11.3 persons per square kilometre, and eight cities with more than 50,000 inhabitants. Oslo is the capital, with about 500,000 inhabitants. Bergen is the second largest town with about 225,000 people, while about 145,000 are living in Trondheim, and 105,000 in Stavanger. Tromsø (the eight largest Norwegian city), which is the largest town up in the north with 58,000 inhabitants was also included.
22. Lien (1987).
23. According to Joanne Finkelstein's proposition, with some adaptation. Finkelstein (1989), 28.
24. This type of food can easily be compared with the typical *kro* food. *Kro* are those eating places along the roads, on which most car tourists are dependent. They usually do not have any local or national food, but mostly serve *løvbiff*, hamburger, frozen pommes frites or possibly frozen salmon.

25. See 'Could the lowly coalfish become haute cuisine' at http://odin.dep.no.
26. Whale becoming trendy food could be seen as a reaction against the international agreement, which states that killing whales for food is banned.
27. Just an anecdotal comparison, with a menu from an exclusive restaurant in Stavanger in 1872, in which food was directly imported from the continental Europe, especially France.
28. See http://odin.dep.no/html/nofovalt/depter/ud/publ/97/nortaste/en/main.html; main address with information about Norway http://odin.dep.no.
29. Cit. in Stø (1999).
30. This is part of a more general attitude towards food. Some recent articles on opinions about food and health, showed that 72 per cent of Norwegian people totally agreed with the sentence "the food I buy is important for my health", and that "trust in experts" was an important factor. Wandel & Fagerli (1999).
31. The semantic meaning of the word traditional depends on how it is defined by people working in restaurants. At this level, it is not possible to discuss what tradition really means, or if this image of tradition is a myth or a reality. Anyway, a collective myth often becomes a reality.
32. *Raspeboller* are a kind of potatoes cakes, *lutefisk* is dried cod prepared in a potash lye and *pinnekjøtt* are smoked and dried lamb cotelettes.
33. Bomann-Larsen (1996). The Norwegian knife hanging at the belt could also have other meanings from its association with the Resistance under the Second World War, but it just confirms the close relationship between peasant world, and the feeling of being a real Norwegian. The aim of the book is otherwise to show that the Norwegian national identity was mostly built around one man, a lover of nature and sport, Fridtjof Nansen.
34. See, for example, the books by Norwegian folklorists as Anne Eriksen, Olav Chritiansen, or anthropologists as Thomas Hylland Eriksen.
35. Flandrin & Montanari (1997).
36. In Norway after the new Constitution of 1814, the few titles of nobility totally disappeared for political reasons.
37. Fischler (1993), 245.
38. Ritzer (1998); see also an excellent article by Wang (1999).

BIBLIOGRAPHY

Ambjørnrud, O. et al. (eds), *Norsk mat* (Oslo, 1965).

Amilien, V., 'Is grandmother's cuisine traditional food? A concept definition in tourism research', paper presented at *8th Nordic Symposium in Hospitality and Tourism Research*, Alta, 18.–21.11.1999. The paper is on Internet at http://www.sifo.no/publikasjon/sammendrag/paper-amilien.htm.

Anon., *Three in Norway, by two of them* (Oslo, 1998, 1st edn 1882).

Bomann-Larsen, Tor, *Den evige sne : en skihistorie om Norge* (Cappelen, 1996).

Bonifacius, C., *Fornuftig madstel* (Christiana, 1864).

Brimi, Arne, *A taste of Norway, a cookbook based on Nature's own ingredients* (Oslo, 1987).

Bugge, A. & Døving, R., *Det norske måltidsmønsteret – ideologi og praksis* (Lysakcr, 2000).

Døving, R., 'Matpakke: Den store norske fortellingen om familien og nasjonen'. *Din* 1 (1999).

Fagerli, R. A., *Endringer i nordmenns matvaner på 80- og 90-tallet*, SIFO Report 1 (1999).

Finkelstein, J., *Dining out* (Oxford, 1989).

Flandrin J-L. & Montanari M. (eds), *L'histoire de l'alimentation* (Paris, 1997).

Fischler, C., *L'homnivore* (Paris, 1993).
Kjærnes, U., *Eating patterns. A day in the lives of Nordic peoples*, SIFO Report 7 (2001).
Lien, M. E. 'Fra bokna fesk til pizza'. *Oslo Occasional Papers in Social Anthropology* 18 (Oslo, 1987).
Lyngø, I. J., 'Vitaminene: "disse vidunderliuge stoffe"'. *Din* 1 (1999).
Maurizio, A., *Histoire de l'alimentation végétale* (Paris, 1932).
Mennell S. 'Divergences and convergences in the development of culinary cultures', in: Hans J. Teuteberg (ed.), *European Food History. A Research Review* (Leicester / London / New York, 1992).
Official Statistics of Norway, Statistics on Travel 1988 (ref. B866) & 1997 (ref. C486).
Ritzer G., *The McDonaldization Thesis: explorations and extensions* (London, 1998).
Stø, E., 'Norwegian consumers' attitudes towards domestic and foreign products', paper presented at *Food quality: regional and national aspects in contemporary Europe*, Stockholm, 26.–28.3.1999.
Wandel, M. & Fagerli, R. A., 'Norwegians' Opinions of a Healthy Diet in Different Stages of Life'. *Journal of Nutrition Education* (1999).
Wang, N., 'Rethinking authenticity in tourism experience'. *Annals of Tourism Research* 26 (2) (1999), 349–370.
Wold, B., *Mat: arbeid og kultur*, SIFO Delrapport 1 (1985).

Index

A

B

C

D

M

N

O

P

V

W

Y

Z